THE AXIOM OF CHOICE

THOMAS J. JECH
Professor Emeritus
The Pennsylvania State University

DOVER PUBLICATIONS, INC.
Garden City, New York

Copyright

Bibliographical Note

This Dover edition, first published in 2008, is an unabridged republication of the work originally published in 1973 by North-Holland Publishing Company, Amsterdam, and American Elsevier Publishing Company, Inc., New York, as Volume 75 in the North-Holland series Studies in Logic and The Foundations of Mathematics.

Library of Congress Cataloging-in-Publication Data

Jech, Thomas J.
The axiom of choice / Thomas J. Jech. — Dover ed.
p. cm.
Originally published: Amsterdam : North-Holland Pub. Co. ; New York : American Elsevier Pub. Co., 1973. Studies in logic and the foundations of mathematics, v. 75.
ISBN-13: 978-0-486-46624-8
ISBN-10: 0-486-46624-8
1. Axiom of choice. I. Title.

QA248.J4 2008
511.3—dc22

2008014612

Manufactured in the United States by LSC Communications
46624807 2023
www.doverpublications.com

To my parents

PREFACE

The book was written in the long Buffalo winter of 1971–72. It is an attempt to show the place of the Axiom of Choice in contemporary mathematics. Most of the material covered in the book deals with independence and relative strength of various weaker versions and consequences of the Axiom of Choice. Also included are some other results that I found relevant to the subject.

The selection of the topics and results is fairly comprehensive, nevertheless it is a selection and as such reflects the personal taste of the author. So does the treatment of the subject. The main tool used throughout the text is Cohen's method of forcing. I tried to use as much similarity between symmetric models of ZF and permutation models of ZFA as possible. The relation between these two methods is described in Chapter 6, where I present the Embedding Theorem (Jech–Sochor) and the Support Theorem (Pincus).

Consistency and independence of the Axiom of Choice (due to Gödel and Cohen, respectively) are presented in Chapters 3 and 5. Chapter 4 introduces the permutation models (Fraenkel, Mostowski, Specker). Chapters 2 and 10 are devoted to the use of the Axiom of Choice in mathematics. Chapter 2 contains examples of proofs using the Axiom of Choice, whereas Chapter 10 provides counterexamples in the absence of the Axiom of Choice.

Chapters 7 and 8 deal with various consequences of the Axiom of Choice. The topics considered in Chapter 7 include the Prime Ideal Theorem, the Ordering Principle and the Axiom of Choice for Finite Sets, while Chapter 8 is concerned with the Countable Axiom of Choice, the Principle of Dependent Choices and related topics. In Chapter 9 we discuss the differences between permutation models of ZFA and symmetric models of ZF.

Chapter 11 is devoted to questions concerning cardinal numbers in set theory without the Axiom of Choice. Finally, in Chapter 12 we discuss briefly the Axiom of Determinacy and related topics. This subject has been recently in the focus of attention, and it seems that solution of the open problems presented at the end of Chapter 12 would mean an important contribution to the investigation of the Foundations of Mathematics.

The book is more or less self-contained. Most of the theorems are accompanied by a proof of a sort. There are, however, several instances where I elected to present a theorem without proof, and an insatiable reader will have to resort to looking up the proof in the journals.

Each chapter contains a number of problems. Some of them are simply exercises, but many of the problems are actually important results which for various reasons are not included in the main text. These are in most cases provided with an outline of proof. To help the reader, I adopted the classification used by producers of French brandy: The one star problems are generally more difficult than the plain problems. Two stars are reserved for results presented without a proof (in most cases, such a result is the main theorem of a long paper). And finally, three stars indicate that the problem is open (there are still a few left).

Among those whose work on the subject has had most influence on this book are (apart from Gödel and Cohen) A. Levy, A. Mostowski and A. Tarski, to name a few. My special thanks are due to P. Vopěnka, who is responsible for my interest in the subject and under whose guidance I wrote my 1966 thesis 'The Axiom of Choice'.

I am very much indebted to D. Pincus, for two reasons. Firstly, many of the results of his 1969 thesis found its way into this book, and secondly, his correspondence with me throughout the past five years had a significant influence on my treatment of the subject. Finally, I am grateful to U. Felgner for his valuable comments after he was so kind to read my manuscript last summer.

Princeton, N.J. December 1972

CONTENTS

CHAPTER 1

INTRODUCTION

1.1. The Axiom of Choice

A mathematician of the present generation hardly considers the use of the Axiom of Choice a questionable method of proof. As a result of algebra and analysis going abstract and the development of new mathematical disciplines such as set theory and topology, practically every mathematician learns about the Axiom of Choice (or at least of its most popular form, Zorn's Lemma) in an undergraduate course. He also probably vaguely knows that there has been some controversy involving the Axiom of Choice, but it has been resolved by the logicians to a general satisfaction.

There has been some controversy about the Axiom of Choice, indeed. First, let us see what the Axiom of Choice says:

AXIOM OF CHOICE. *For every family $\mathscr{F}$ of nonempty sets, there is a function f such that $f(S) \in S$ for each set S in the family $\mathscr{F}$.*

(We say that f is a *choice function* on $\mathscr{F}$.) In some particular instances, this is self-evident. E.g., let $\mathscr{F}$ consist of sets of the form $\{a, b\}$, where a and b are real numbers. Then the function

$$f(\{a, b\}) = \min(a, b)$$

is a choice function on $\mathscr{F}$. Or, if all the sets in $\mathscr{F}$ are *singletons*, i.e. sets of the form $\{a\}$, then one can easily find a choice function on $\mathscr{F}$. Finally, if $\mathscr{F}$ is a *finite* family of nonempty sets, then there is a choice function. (To show this, one uses induction on the size of $\mathscr{F}$; naturally, a choice function exists for a family which consists of a single nonempty set.)

Thus the idea of having a choice function on *every* family of nonempty sets is not against our mathematical mind. But try to think of a choice function on the family of *all* nonempty sets of reals! Or, let $\mathscr{F}$ be the family of all (unordered) *pairs* $\{S, T\}$, where S and T are *sets of reals*.

After thinking about these problems for a while, one should start to appreciate the other point of view. The Axiom of Choice *is* different from the ordinary principles accepted by mathematicians. And this was one of the sources of objections to the Axiom of Choice, as late as in the thirties. The other source of objections is the fact that the Axiom of Choice can be used to prove theorems which are to a certain extent 'unpleasant', and even theorems which are not exactly in line with our 'common sense' intuition.

The rest of Chapter 1 is devoted to such examples.

1.2. A nonmeasurable set of real numbers

The most popular example of an 'unpleasant' consequence of the Axiom of Choice is the existence of a set of real numbers which is not Lebesgue measurable.

Let $\mu(X)$ denote the Lebesgue measure of a set X of real numbers. As is well-known, μ is countably additive, translation invariant and $\mu([a, b]) = b-a$ for every interval $[a, b]$.

For real numbers in the interval $[0, 1]$, let us define

$$x \sim y$$

whenever $x-y$ is a rational number. The relation $\sim$ on $[0, 1]$ is an equivalence; denote by $[x]$ the corresponding equivalence class, for each $x \in [0, 1]$. By the Axiom of Choice, we can choose one element out of each equivalence class. Thus there is a set M of real numbers, $M \subset [0, 1]$, with the property that for each real number x, there exists a unique $y \in M$ and a unique rational number r such that $x = y+r$. If we let

$$M_r = \{y+r : y \in M\}$$

for each rational number r, then we have a partition of the set R of all real numbers into countably many disjoint sets

$$R = \bigcup \{M_r : r \text{ rational}\}. \tag{1.1}$$

It follows that the set M is not measurable. Assuming that M is measurable, we get a contradiction as follows: First, $\mu(M) = 0$ is impossible, since this would imply $\mu(R) = 0$, using (1.1). However, $\mu(M) > 0$ is also impossible, since this would imply

$$\mu([0, 1]) \geqslant \mu(\bigcup \{M_r : r \text{ rational and } 0 \leqslant r \leqslant 1\}) = \sum_{\substack{0 \leqslant r \leqslant 1 \\ r \text{ rational}}} \mu(M_r) = \infty,$$

using the fact that each M_r would have to have the same measure as M.

1.3. A paradoxical decomposition of the sphere

The theorems presented here demonstrate what 'absurd' results can be expected when employing the Axiom of Choice.

THEOREM 1.1. *A sphere S can be decomposed into disjoint sets*

$$S = A \cup B \cup C \cup Q$$

such that:

(i) *the sets A, B, C are congruent to each other;*
(ii) *the set $B \cup C$ is congruent to each of the sets A, B, C;*
(iii) *Q is countable.*

Let us consider the following relation between sets in the three-dimensional Euclidean space: We let

(1.2) $$X \approx Y$$

if there is a finite decomposition of X into disjoint sets

$$X = X_1 \cup \ldots \cup X_m$$

and a decomposition of Y into the same number of disjoint sets

$$Y = Y_1 \cup \ldots \cup Y_m$$

such that X_i is congruent to Y_i for each $i = 1, \ldots, m$. We have the following remarkable theorem:

THEOREM 1.2. *A closed ball U can be decomposed into two disjoint sets,*

$$U = X \cup Y$$

such that $U \approx X$ and $U \approx Y$.

Thus, using the Axiom of Choice, one can cut a ball into a finite number of pieces and rearrange them to get two balls of the same size as the original one!

The following sequence of lemmas will establish the proofs of Theorems 1 and 2. Let G be the free product of the groups $\{1, \varphi\}$ and $\{1, \psi, \psi^2\}$, i.e., the group of all formal products formed by φ, ψ and ψ^2, with the specification that $\varphi^2 = 1$ and $\psi^3 = 1$. Let us consider two axes of rotation a_φ, a_ψ going through the center of the ball U, and consider the group of all rotations generated by a rotation φ by 180° about a_φ and a rotation ψ by 120° about a_ψ.

LEMMA 1.3. *We can determine the axes a_φ and a_ψ in such a way that distinct elements of G represent distinct rotations generated by φ and ψ.*

PROOF (outline). What we have to do is to determine the angle ϑ between a_φ and a_ψ in such a way that no element of G other than 1 represents the identity rotation. Consider a typical element of G, e.g. a product of the type

$$\alpha = \varphi \cdot \psi^{\pm 1} \dots \cdot \varphi \cdot \psi^{\pm 1}. \tag{1.3}$$

Using the equations for orthogonal transformations and some elementary trigonometry, one can prove that the equation

$$\alpha = 1,$$

where α is some product of type (1.3), has only finitely many solutions. Consequently, except for a countable set of values, we are free to select any angle ϑ to satisfy the requirements.

Thus we consider such ϑ and then G is the group of all rotations generated by φ and ψ.

LEMMA 1.4. *The group G can be decomposed into three disjoint sets*

$$G = A \cup B \cup C$$

such that

$$A \cdot \varphi = B \cup C, \qquad A \cdot \psi = B, \qquad A \cdot \psi^2 = C. \tag{1.4}$$

PROOF. We construct A, B, C by recursion on the lengths of the elements of G. We let $1 \in A$, $\varphi, \psi \in B$ and $\psi^2 \in C$ and then continue as follows for any $\alpha \in G$:

		$\alpha \in A$	$\alpha \in B$	$\alpha \in C$
α ends with	$\psi^{\pm 1}$:	$\alpha\varphi \in B$	$\alpha\varphi \in A$	$\alpha\varphi \in A$
	φ:	$\alpha\psi \in B$	$\alpha\psi \in C$	$\alpha\psi \in A$
		$\alpha\psi^{-1} \in C$	$\alpha\psi^{-1} \in A$	$\alpha\psi^{-1} \in B$

This will guarantee that condition (1.4) is satisfied.

So far, we have not used the Axiom of Choice. To complete the proof of Theorem 1.1, we use a similar argument to that in the construction of a nonmeasurable set of real numbers in section 1.2.

Let Q be the set of all fixed points on the sphere S of all rotations $\alpha \in G$. Each rotation $\alpha \in G$ has two fixed points; thus Q is countable. The set $S-Q$ is a disjoint union of all orbits P_x of the group G:

$$P_x = \{x\alpha : \alpha \in G\}.$$

By the Axiom of Choice, there exists a set M which contains exactly one element in each P_x, $x \in S-Q$. If we let

$$A = M \cdot \mathscr{A}, \qquad B = M \cdot \mathscr{B}, \qquad C = M \cdot \mathscr{C},$$

it follows from (1.4) that A, B, C are disjoint, congruent to each other, and $B \cup C$ is congruent to each of them; moreover,

$$S = A \cup B \cup C \cup Q.$$

This completes the proof of Theorem 1.1.

LEMMA 1.5. *Let* $\approx$ *be the relation defined in* (1.2). *Then*:

(a). $\approx$ *is an equivalence.*

(b). *If* X *and* Y *are disjoint unions of* X_1, X_2 *and* Y_1, Y_2, *respectively, and if* $X_i \approx Y_i$ *for each* $i = 1, 2$, *then* $X \approx Y$.

(c). *If* $X_1 \subseteq Y \subseteq X$ *and if* $X \approx X_1$, *then* $X \approx Y$.

PROOF. (a) and (b) are easy.

To prove (c), let $X = X^1 \cup \ldots \cup X^n$ and $X_1 = X_1^1 \cup \ldots \cup X_1^n$ such that X^i is congruent to X_1^i for each $i = 1, \ldots, n$. Choose a congruence $f^i : X^i \to X_1^i$ for each $i = 1, \ldots, n$, and let f be the one-to-one mapping of X onto X^1 which agrees with f^i on each X^i. Now let

$$\begin{aligned} X_0 &= X, & X_1 &= f''X_0, & X_2 &= f''X_1, \ldots, \\ Y_0 &= Y, & Y_1 &= f''Y_0, & Y_2 &= f''Y_1, \ldots, \end{aligned}$$

(where $f''X = \{f(x) : x \in X\}$ is the image of X). If we let

$$Z = \bigcup_{n=0}^{\infty} (X_n - Y_n),$$

then $f''Z$ and $X-Z$ are disjoint, $Z \approx f''Z$ and

$$X = Z \cup (X-Z), \qquad Y = f''Z \cup (X-Z),$$

and hence $X \approx Y$ by (b).

Proof of Theorem 1.2. Now, let U be a closed ball and let

$$S = A \cup B \cup C \cup Q$$

be the decomposition of its surface from Theorem 1. We have

$$U = \bar{A} \cup \bar{B} \cup \bar{C} \cup \bar{Q} \cup \{c\},$$

where c is the centre of the ball and for each $X \subset S, \bar{X}$ is the set of all $x \in U$, other than c, such that its projection onto the surface is in X. Clearly,

$$\bar{A} \approx \bar{B} \approx \bar{C} \approx \bar{B} \cup \bar{C}. \tag{1.5}$$

Let

$$X = \bar{A} \cup \bar{Q} \cup \{c\}, \qquad Y = U - X.$$

From (1.5) and Lemma 1.5, we get

$$\bar{A} \approx \bar{A} \cup \bar{B} \cup \bar{C},$$

and so

$$X \approx U. \tag{1.6}$$

Now, it is easy to find some rotation α (not in G) such that Q and $Q \cdot \alpha$ are disjoint, and so, using

$$\bar{C} \approx \bar{A} \cup \bar{B} \cup \bar{C},$$

there exists $S \subset C$ such that $\bar{S} \approx \bar{Q}$. Let p be some point in $\bar{S} - \bar{C}$. Obviously,

$$\bar{A} \cup \bar{Q} \cup \{c\} \approx \bar{B} \cup \bar{S} \cup \{p\}. \tag{1.7}$$

Since

$$\bar{B} \cup \bar{S} \cup \{p\} \subseteq Y \subseteq U,$$

we can use (1.6), (1.7) and Lemma 1.5(c), to get

$$Y \approx U.$$

This completes the proof of Theorem 1.2.

1.4. Problems

1. Prove that the Axiom of Choice is equivalent to the following principle: Let $\mathscr{F}$ be a family of disjoint nonempty sets. Then there is a set M such that $M \cap X$ contains exactly one element, for each $X \in \mathscr{F}$.

2. Let $\mathscr{F}$ be a family of nonempty sets of natural numbers. Then there exists a choice function on $\mathscr{F}$.

3. Let $\mathscr{F}$ be a family of finite nonempty sets of real numbers. Then there exists a choice function on $\mathscr{F}$.

4. Assuming that there is a nonmeasurable set of real numbers, construct a subset of the interval $[0, 1]$, whose inner measure is 0 and whose outer measure is 1.

A set of real numbers is called *meager* if it is the union of a countable family of nowhere dense sets. A set X has the *Baire property* if for some open set G, the symmetric difference $X \triangle G$ is meager.

5. Without using the Axiom of Choice, prove that the set R of all real numbers is not meager (Baire Category Theorem).

[*Hint*: The intersection of countably many open dense sets is non-empty.]

6. Using the Axiom of Choice, prove that the union of countably many countable sets is countable.

7. Using the Axiom of Choice, prove that the union of countably many meager sets is meager.

8. Assuming that there is a set of real numbers which does not have the Baire property, construct a set X which is not meager and such that for no nonempty open set G, $G-X$ is meager.

9. The set M from Section 1.2 does not have the Baire property.

[*Hint*: If there is an open interval (a, b) such that $(a, b) = M \cup N$, where N is meager, then there is $x \in M$ such that $x+r \notin N$ for any rational r, a contradiction.]

*10. Prove the existence of a nonmeasurable set using only the assumption that a choice function exists on every family of pairs.

[*Hint*[1]: Work in the Cantor space (the set ${}^{\omega}2$ of all functions $a : \omega \to \{0,1\}$) with the product measure. Let $a \sim b$ iff $a(n) \neq b(n)$ for only finitely many n's; let $[a]$ be the equivalence classes of a. Let a^* be such that $a^*(n) = 1$ iff $a(n) = 0$. Using choice for the family of all pairs $\{[a], [a^*]\}$, obtain $A \subset {}^{\omega}2$ such that for each x,

[1] Note that in fact one uses a weaker assumption: There exists a nontrivial ultrafilter over ω.

(i) $x \in A$ if and only if $x^* \notin A$;

(ii) if $x \in A$ and $x \sim y$ then $y \in A$.

By (ii), if some basic interval is included in A up to a set of measure 0, then every basic interval of the same length is; thus the measure of A would have to be either 0 or 1. On the other hand, A should have the same measure as $A^* = \{x^*: x \in A\}$, but $A \cup A^* = {}^{\omega}2$, a contradiction.]

*11. Prove the existence of a set without the Baire property using the same assumption as in Problem 10.

[*Hint*: The set A would have to be either meager or a complement of a meager set.]

1.5. Historical remarks

The Axiom of Choice was first formulated by Zermelo [1904]. The formulation in Problem 1 is due to Russel [1906]. The construction of a nonmeasurable set in Section 1.2 is due to Vitali [1905]; the construction in Problems 10 and 11 follows Sierpiński [1938]. Theorem 1.1 is in Hausdorff [1914], and Theorem 1.2 was proved by Banach and Tarski [1924].

CHAPTER 2

USE OF THE AXIOM OF CHOICE

2.1. Equivalents of the Axiom of Choice

While in the first chapter we tried to convince the reader that the Axiom of Choice has unpleasant consequences, we shall devote this chapter to the task of improving its image by presenting several important theorems of contemporary mathematics in which the Axiom of Choice is indispensable.

We start by giving several equivalents of the Axiom of Choice. The literature gives numerous examples of theorems which are equivalent to the Axiom of Choice. Most popular among them are the three following principles.

WELL-ORDERING PRINCIPLE (Zermelo's Theorem). *Every set can be well-ordered.*

We recall that an ordering $<$ of a set S is a *well-ordering* if every nonempty $X \subseteq S$ has a least element in the ordering $<$.

To state the second principle, we need some definitions: A subset C of a partially ordered set $(P, <)$ is a *chain* in P if C is linearly ordered by $<$; u is an *upper bound* of C if $c \leqslant u$ for each $c \in C$. We say that $a \in P$ is a *maximal element* if $a < x$ for no $x \in P$.

MAXIMAL PRINCIPLE I (Zorn's Lemma). *Let $(P, <)$ be a nonempty partially ordered set and let every chain in P have an upper bound. Then P has a maximal element.*

Finally, let $\mathscr{F}$ be a family of sets. We say that $\mathscr{F}$ has *finite character* if for each X, X belongs to $\mathscr{F}$ if and only if every finite subset of X belongs to $\mathscr{F}$.

MAXIMAL PRINCIPLE II (Tukey's Lemma). *Let $\mathscr{F}$ be a nonempty family of sets. If $\mathscr{F}$ has finite character, then $\mathscr{F}$ has a maximal element (maximal with respect to inclusion $\subseteq$).*

THEOREM 2.1. *The following statements are equivalent:*

(i) *Axiom of Choice.*
(ii) *Well-ordering Principle.*
(iii) *Maximal Principle* I.
(iv) *Maximal Principle* II.

PROOF. (i) $\Rightarrow$ (ii). Let S be a set. To find a well-ordering of S means to find an ordinal number α and a one-to-one α-sequence

$$a_0, a_1, \ldots, a_\xi, \ldots \qquad (\xi < \alpha) \tag{2.1}$$

which enumerates S. Let F be a choice function on the family of all nonempty subsets of S; we construct a sequence (2.1) by transfinite recursion:

$$\begin{aligned} a_0 &= F(S), \\ a_\xi &= F(S - \{a_\eta : \eta < \xi\}). \end{aligned}$$

The construction stops as soon as we exhaust all elements of S.

(ii) $\Rightarrow$ (iii). Let $(P, <)$ be a nonempty partially ordered set and assume that every chain in P has an upper bound. We find a maximal element of P. By assumption, the set P can be well-ordered, i.e., there is an enumeration

$$P = \{p_0, p_1, \ldots, p_\xi, \ldots\} \qquad (\xi < \alpha)$$

for some ordinal number α. By transfinite recursion, let

$$c_0 = p_0$$

and

$$c_\xi = p_\gamma,$$

where γ is the least ordinal such that p_γ is an upper bound of the chain $C = \{c_\eta : \eta < \xi\}$ and $p_\gamma \notin C$. Note that $\{c_\eta : \eta < \xi\}$ is always a chain and that p_γ exists unless $c_{\xi-1}$ is a maximal element of P. Eventually, the construction comes to a halt, and we obtain a maximal element of P.

(iii) $\Rightarrow$ (iv). Let $\mathscr{F}$ be a nonempty family of sets and assume that $\mathscr{F}$ has finite character. $\mathscr{F}$ is partially ordered by the inclusion $\subseteq$. If $\mathscr{C}$ is a chain in $\mathscr{F}$ and if $A = \bigcup \{X : X \in \mathscr{C}\}$, then every finite subset of A belongs to $\mathscr{F}$ and therefore A belongs to $\mathscr{F}$. Obviously, A is an upper bound of $\mathscr{C}$. Hence we may apply Zorn's Lemma and get a maximal element of the family $\mathscr{F}$.

(iv) ⇒ (i). Let $\mathscr{F}$ be a family of nonempty sets. We want to find a choice function on $\mathscr{F}$. Let us consider the family

$$\mathscr{G} = \{f : f \text{ is a choice function on some } \mathscr{E} \subseteq \mathscr{F}\}.$$

A subset of a choice function is a choice function and thus it is easy to see that $\mathscr{G}$ has finite character. By assumption, $\mathscr{G}$ has a maximal element F. A simple appeal to the maximality of F shows that the domain of F is the whole family $\mathscr{F}$.

2.2. Some applications of the Axiom of Choice in mathematics

The equivalents of the Axiom of Choice which were discussed in Section 2.1 are the most popular variants of the choice principle and are most often used in mathematical proofs. In this section, we give several examples of mathematical theorems whose proofs require the use of one of the forms of the Axiom of Choice.

2.2.1. *Example*: *Tychonoff Product Theorem*

Let $\{S_i : i \in I\}$ be a family of topological spaces. The *product space*

$$S = \times\{S_i : i \in I\}$$

is defined as the Cartesian product of the family $\{S_i : i \in I\}$ endowed with the *product topology*.

TYCHONOFF'S THEOREM. *The product of a family of compact spaces is compact in the product topology.*

One popular proof of this theorem uses the characterization of compact spaces by filters.

A *filter* $\mathscr{F}$ over a set S is a nonempty proper subset of the family of all subsets of S such that

(i) $X \in \mathscr{F}$ and $X \subseteq Y$ implies $Y \in \mathscr{F}$;

(ii) $X \in \mathscr{F}$ and $Y \in \mathscr{F}$ implies $X \cap Y \in \mathscr{F}$.

A filter is an *ultrafilter* if

(iii) for each $X \subseteq S$, either $X \in \mathscr{F}$ or $S - X \in \mathscr{F}$.

A *filter-base* is a family $\mathscr{B}$ such that

$$\{X \supseteq X_1 \cap \ldots \cap X_k : X_1, \ldots, X_k \in \mathscr{B}\}$$

is a filter. A space S is *compact* if for every filter-base $\mathscr{B}$ over S,

$$\bigcap \{\overline{X} : X \in \mathscr{B}\} \tag{2.2}$$

is nonempty.

Let

$$S = \mathsf{X}\{S_i : i \in I\}$$

be a product of compact spaces. The family of all filter-bases over S has finite character and so by Tukey's Lemma (or rather by its self-refinement, cf. Problem 3), every filter-base is included in a maximal filter-base (an ultrafilter). To show that the intersection (2.2) is nonempty whenever $\mathscr{B}$ is maximal (which is obviously sufficient), we use:

(1) the compactness of the spaces S_i to show that the intersections of the projections $A_i = \bigcap \{\overline{\mathrm{pr}_i(X)} : X \in \mathscr{B}\}$ are nonempty;

(2) the Axiom of Choice again, to pick x_i in each A_i, and

(3) the maximality of $\mathscr{B}$ to show that the point $x = \langle x_i : i \in I\rangle$ belongs to the intersection (2.2).

2.2.2. *Example*

THEOREM. *Every vector space has a basis.*

PROOF. Let $\mathscr{F}$ be the family of all linearly independent sets of vectors. Obviously, $\mathscr{F}$ has finite character and so by Tukey's Lemma, there exists a maximal linearly independent set B. Using the maximality, one proves without much effort that B is a basis.

2.2.3. *Example*: *Nielsen–Schreier Theorem*

A group G is a *free group* if it has a set of generators A with the following property: Every element g of G other than 1 can be *uniquely* written in the form

$$a_1^{\pm 1} a_2^{\pm 1} \ldots a_n^{\pm 1},$$

where the a_i are in A and no a appears adjacent to an a^{-1}.

NIELSEN–SCHREIER THEOREM. *Every subgroup of a free group is a free group.*

The proof of this theorem is rather involved and we will not present it. Let us only mention that the standard proof uses the fact that the given subgroup can be well-ordered and this is used to construct, by transfinite recursion, a set of free generators of the subgroup.

2.2.4. *Example*: *Hahn–Banach Theorem*

Let E be a real vector space. A functional p on E is *sublinear* if $p(x+y) \leqslant p(x)+p(y)$ for all $x, y \in E$ and $p(rx) = rp(x)$ for all nonnegative reals r and all $x \in E$.

HAHN–BANACH EXTENSION THEOREM. *Let p be a sublinear functional on E and let φ be a linear functional defined on a subspace V of E such that $\varphi(x) \leqslant p(x)$ on V. Then there exists a linear functional ψ on E which extends φ, and $\psi(x) \leqslant p(x)$ holds for all $x \in E$.*

PROOF. One considers the family $\mathscr{F}$ of all linear functionals ψ extending φ, defined on a subspace of E, and satisfying $\psi(x) \leqslant p(x)$. The family $\mathscr{F}$, partially ordered by extension, satisfies the assumption of Zorn's Lemma and so there is a maximal functional ψ in $\mathscr{F}$. It remains to be verified that a maximal ψ is defined everywhere on E. This requires a lemma to the effect that if $\psi \in \mathscr{F}$ and $x \in E$ is outside the domain of ψ, then there exists a $\bar{\psi} \in \mathscr{F}$ which is defined at x.

It should be mentioned that the Hahn–Banach Theorem can be proved from a weaker assumption than the full Axiom of Choice. In Problem 19, the reader will find an equivalent of the Hahn–Banach Theorem, and its relative strength will be discussed in due course.

2.2.5. *Example: Algebraic closure of a field*

An *algebraic closure* of a field F is an algebraically closed extension C of F which is algebraic over F; i.e., every nonconstant polynomial has a root and every element of C is a root of a polynomial in $F[x]$.

THEOREM. *For any field F, an algebraic closure of F exists and is unique (up to isomorphism).*

PROOF (outline): For every polynomial $p \in F[x]$ of nth degree, consider n indeterminates $y_1^{(p)}, \ldots, y_n^{(p)}$; let Y be the set of all such $y_i^{(p)}$. In the ring $F[x][Y]$, let I be the ideal generated by all the elements of the form

$$p(x) - (x - y_1^{(p)}) \cdot \ldots \cdot (x - y_n^{(p)}).$$

Using Zorn's Lemma, there is a maximal ideal $J \supseteq I$ in the ring $F[x][Y]$. The quotient $F[x][Y]/J$ is a field, extends F and is algebraically closed. Once we have one algebraically closed extension of F, we can get one which is algebraic over F. The uniqueness of algebraic closure follows by application of Zorn's Lemma to the family of all partial isomorphisms between the two given algebraic closures of F.

2.2.6. *Example: Maximal Ideal Theorem for Lattices*

Let L be a lattice. An *ideal* in L is a nonempty proper subset I of L such that

(i) $a \in I$ and $b \leqslant a$ implies $b \in I$;

(ii) $a \in I$ and $b \in I$ implies $a \vee b \in I$.

MAXIMAL IDEAL THEOREM FOR LATTICES. *Every lattice with unit and at least one other element has a maximal ideal.*

PROOF. Let $\mathscr{F}$ be the family of all ideals in the given lattice. The family $\mathscr{F}$ satisfies the assumption of Zorn's Lemma and so it has a maximal element.

2.3. The Prime Ideal Theorem

In the examples of the preceding section, we showed that using the Axiom of Choice, we can establish the existence of maximal ideals in rings, lattices, or set algebras. Among the theorems of this type, one – the (*Boolean*) *Prime Ideal Theorem* – plays a particularly prominent role. Firstly, because it can be used in many proofs instead of the Axiom of Choice; secondly, because it is equivalent to several other statements; and finally, because it is essentially weaker than the general Axiom of Choice (this will be discussed in Chapter 7).

A *Boolean algebra* is an algebra B with two binary operations $+$, $\cdot$ (Boolean *sum* and *product*), one unary operation $-$ (*complement*) and two constants 1, 0. The axioms governing Boolean algebras are the following:

$$u+u = u, \qquad u \cdot u = u,$$
$$u+v = v+u, \qquad u \cdot v = v \cdot u,$$
$$u+(v+w) = (u+v)+w, \qquad u \cdot (v \cdot w) = (u \cdot v) \cdot w,$$
$$(u+v) \cdot w = (u \cdot w)+(v \cdot w),$$
$$(u \cdot v)+w = (u+w) \cdot (v+w),$$
$$u+(-u) = 1, \qquad u \cdot -u = 0,$$
$$-(u+v) = -u \cdot -v, \qquad -(u \cdot v) = -u+-v.$$
$$- -u = u.$$

Besides the above operations, one often considers a partial ordering $\leqslant$ of B which is introduced in terms of $+$ by

$$u \leqslant v \leftrightarrow u+v = v.$$

Then 1 and 0 are the greatest and the smallest element of B, respectively, and $+$ and $\cdot$ represent the l.u.b. and the g.l.b., respectively.

An *ideal* I on B is a nonempty proper subset of B such that:

(i) $u \in I$ and $v \leqslant u$ implies $v \in I$;

(ii) $u \in I$ and $v \in I$ implies $u+v \in I$.

An ideal is a *prime ideal* if

(iii) for each $u \in B$, either $u \in I$ or $-u \in I$.

In a Boolean algebra, an ideal is a prime ideal if and only if it is maximal.

The dual notions of an ideal and a prime ideal are a *filter* and an *ultrafilter* on B, satisfying:

(i) $u \in F$ and $v \geqslant u$ implies $v \in F$;

(ii) $u \in F$ and $v \in F$ implies $u \cdot v \in F$;

and for an ultrafilter:

(iii) for each $u \in B$, either $u \in F$ or $-u \in F$.

PRIME IDEAL THEOREM. *Every Boolean algebra has a prime ideal.*

The Prime Ideal Theorem is a consequence of the Axiom of Choice; the proof is an elementary exercise in the use of Zorn's Lemma. It will be shown in a later chapter that the converse is not true. The Axiom of Choice cannot be proved from the Prime Ideal Theorem (in contrast to that, see Problems 10,11). We shall give several equivalents of the Prime Ideal Theorem. First, note that the Prime Ideal Theorem is equivalent to its stronger version:

In every Boolean algebra, every ideal can be extended to a prime ideal.

To show that the stronger version follows from the weaker version, let B be a Boolean algebra and I an ideal in B. Consider the equivalence relation

$$u \sim v \leftrightarrow (u \cdot -v) + (v \cdot -u) \in I.$$

Let C be the set of all equivalence classes $[u]$ and define operations $+$, $\cdot$ and $-$ on C as follows:

$$[u] + [v] = [u+v], \qquad [u] \cdot [v] = [u \cdot v], \qquad -[u] = [-u].$$

Then C is a Boolean algebra, the *quotient* of B:

$$C = B/I.$$

By the Prime Ideal Theorem, C has a prime ideal K. It is easy to verify that the set

$$J = \{u \in B : [u] \in K\}$$

is a prime ideal in B and $I \subseteq J$.

Using the duality between ideals and filters, we have the following formulations of the Prime Ideal Theorem:

Every Boolean algebra has an ultrafilter,

or:

Every filter in a Boolean algebra can be extended to an ultrafilter.

In the special case that the algebra is the set $\mathscr{P}(S)$ of all subsets of S, with set theoretical operations $\cup$, $\cap$ and $-$; the notion of a filter and ultrafilter coincides with the earlier definition. Thus we have:

ULTRAFILTER THEOREM. *Every filter over a set S can be extended to an ultra-filter.*

We shall show that the Ultrafilter Theorem is equivalent to the Prime Ideal Theorem. However, the corresponding weaker version of the Ultra-filter Theorem, viz.

For each nonempty set S, there is an ultrafilter over S,

is trivially true without any use of choice: Let $p \in S$; then the set

$$\{X \subseteq S : p \in X\}$$

is a *trivial* ultrafilter over S. Let us say here that even the statement

For each infinite set S, there is a nontrivial ultrafilter over S,

although unprovable without the Axiom of Choice, is essentially weaker than the Prime Ideal Theorem (see Problem 8.5).

Another principle to be considered here is the following: Let S be a set and let M be a set of functions defined on finite subsets of S, with values 0 or 1. We say that M is a *binary mess* on S if M satisfies the following properties:

(i) For each finite $P \subseteq S$, there is $t \in M$ such that t is defined on P; $\mathrm{dom}\,(t) = P$.

(ii) For each $t \in M$ and each finite $P \subseteq S$, the restriction $t|P$ belongs to M.

Let f be a function on S with values 0, 1. Then f is *consistent* with a mess M if every restriction $f|P$, P a finite subset of S, belongs to M.

CONSISTENCY PRINCIPLE. *For every binary mess M on a set S, there exists a function f on S which is consistent with M.*

The last principle to be mentioned here is the Compactness Theorem for the first order predicate logic. Let $\mathscr{L}$ be a language of first-order logic; the set of predicates in $\mathscr{L}$ is arbitrary, not necessarily countable. Let Σ be a set of sentences of $\mathscr{L}$.

COMPACTNESS THEOREM. *If every finite subset of Σ has a model, then Σ has a model.*

We shall show that the above principles are all equivalent to the Prime Ideal Theorem.

THEOREM 2.2. *The following are equivalent:*

(i) *Prime Ideal Theorem.*
(ii) *Ultrafilter Theorem.*
(iii) *Consistency Theorem.*
(iv) *Compactness Theorem.*

PROOF. (i) ⇒ (ii). The Ultrafilter Theorem is an immediate consequence of the stronger version of the Prime Ideal Theorem.

(ii) ⇒ (iii). Let M be a binary mess on S; we shall find a binary function f on S, that is consistent with M. Let I be the set of all finite subsets P of S. For each $P \in I$, let M_P be the finite set of all $t \in M$ such that $\operatorname{dom}(t) = P$. Let Z be the set of all functions z such that:

(a) $\operatorname{dom}(z) \subseteq I$;
(b) $z(P) \in M_P$ for each $P \in \operatorname{dom}(z)$;
(c) for any $P, Q \in \operatorname{dom}(z)$, the functions $t_1 = z(P)$ and $t_2 = z(Q)$ are compatible.

Let F be the filter over Z generated by the sets

$$X_P = \{z \in Z : P \in \operatorname{dom}(z)\}$$

(the set $\{X_P : P \in I\}$ is a filter-base). By the Ultrafilter Theorem, there is an ultrafilter U over Z which extends F. For every $P \in I$, the set X_P is a finite disjoint union

$$X_P = X_{t_1} \cup \ldots \cup X_{t_m},$$

where $\{t_1, \ldots, t_m\} = M_P$ and

$$X_t = \{z \in Z : z(P) = t\}.$$

Thus there is a unique $t = t(P) \in M_P$ such that $X_t \in U$. The functions t_P, $P \in I$, are pairwise compatible and thus their union is a binary function f on S which is consistent with M.

(iii) ⇒ (iv). Let Σ be a consistent set of sentences of a first-order language $\mathscr{L}$. By a well-known labeling method, every consistent theory can be expanded to a consistent Skolem theory and so we may assume that Σ contains all the Skolem sentences σ_ϕ for all formulas ϕ of the language $\mathscr{L}$ with one free variable; i.e., σ is the sentence

$$\exists v\, \phi(v) \rightarrow \phi(e_\phi),$$

where e_ϕ is the Skolem constant associated with ϕ. It is sufficient to find an extension Σ^* of Σ which is complete; then one can build a model of Σ^* from the constant terms. Thus the problem is to embed a consistent theory into a complete theory.

Let S be the set of all sentences of the language $\mathscr{L}$. We define a mess M on S as follows:

$$t \in M \leftrightarrow (\exists \mathfrak{A})[\mathfrak{A} \text{ is a model of } \Sigma \cap \operatorname{dom}(t) \text{ and } \forall \sigma \in \operatorname{dom}(t)\ (t(\sigma) = 1 \leftrightarrow \mathfrak{A} \models \sigma)].$$

Since every finite subset of Σ has a model, M is a mess. By the Consistency Principle, there exists a binary function f on S, consistent with M. Now we let

$$\Sigma^* = \{\sigma \in S : f(\sigma) = 1\}.$$

Σ^* is a complete theory and $\Sigma \subseteq \Sigma^*$.

(iv) $\Rightarrow$ (i). Let B be a Boolean algebra. Let $\mathscr{L}$ be a language which has a constant for every element of B. (We can identify $u \in B$ with the corresponding constant in $\mathscr{L}$.) Furthermore, let $\mathscr{L}$ contain a unary predicate I. Let Σ be the following set of sentences of $\mathscr{L}$:

$$I(0), \qquad \neg I(1),$$
$$I(u) \vee I(-u) \quad (\text{for each } u \in B),$$
$$I(u_1) \wedge \ldots \wedge I(u_k) \rightarrow I(u_1 + \ldots + u_k) \ \ (\text{for all } u_1, \ldots, u_k \in B).$$

Every finite subset of Σ has a model. To see that, realize that every finite subset of B generates a finite subalgebra of B and that every finite Boolean algebra has a prime ideal. By the Compactness Theorem, Σ has a model. This in turn yields a prime ideal on B.

Now we shall give some examples of the use of the Prime Ideal Theorem in mathematical proofs. More examples will be given in the problem section.

2.3.1. *Example: Stone Representation Theorem*

A family $\mathscr{A}$ of subsets of a given set S is a *set-algebra* if:

(2.3) (a) $S \in \mathscr{A}$,

(b) if $X, Y \in \mathscr{A}$ then $X \cap Y \in \mathscr{A}$ and $X \cup Y \in \mathscr{A}$,

(c) if $X \in \mathscr{A}$ then $S - X \in \mathscr{A}$.

STONE REPRESENTATION THEOREM. *Every Boolean algebra is isomorphic to a set algebra.*

PROOF. Let B be a Boolean algebra; let

$$S = \{U : U \text{ is an ultrafilter on } B\}.$$

For $u \in B$, we let

$$\pi(u) = \{U \in S : u \in U\}.$$

It is easy to see that

$$\pi(u+v) = \pi(u) \cup \pi(v),$$
$$\pi(u \cdot v) = \pi(u) \cap \pi(v),$$
$$\pi(-u) = S - \pi.$$

One has to use the Prime Ideal Theorem to show that π is one-to-one. Thus π is an isomorphism of B onto $\mathscr{A} = \pi''B$.

Actually, the Stone Representation Theorem is equivalent to the Prime Ideal Theorem; see Problem 12.

2.3.2. *Example*: *Ordering Principle*

We shall show that the Prime Ideal Theorem implies that every set can be linearly ordered. Actually, we get a stronger theorem. If $<$ and $\prec$ are partial orderings of a set P, then we say that $\prec$ extends $<$ if for any $p, q \in P$, $p < q$ implies $p \prec q$.

ORDER EXTENSION PRINCIPLE. *Every partial ordering of a set P can be extended to a linear ordering of P.*

PROOF. We use the Compactness Theorem. Let $(P, \leqslant)$ be a partially ordered set. Let $\mathscr{L}$ be a language containing constants for all $p \in P$ and a binary predicate $\preccurlyeq$. Let Σ be the set of sentences

$$p \preccurlyeq q \wedge q \preccurlyeq r \rightarrow p \preccurlyeq r \qquad (\text{all } p, q, r \in P),$$
$$p \preccurlyeq q \wedge q \preccurlyeq p \rightarrow p = q \qquad (\text{all } p, q \in P),$$
$$p \preccurlyeq q \vee q \preccurlyeq p \qquad (\text{all } p, q \in P),$$
$$p \preccurlyeq q \qquad (\text{all } p, q \in P \text{ such that } p \leqslant q).$$

Since every finite subset of Σ has a model, the application of the Compactness Theorem yields a linear ordering $\preccurlyeq$ of P which extends $\leqslant$.

2.3.3. *Example*: *Artin–Schreier Theorem*

A field F is an *ordered field* if there is a linear ordering $\leqslant$ of F satisfying:

(2.4) (a) if $a \leqslant b$, then $a+c \leqslant b+c$ for all c;
(b) if $a \leqslant b$ and $c \geqslant 0$, then $a \cdot c \leqslant b \cdot c$.

ARTIN–SCHREIER THEOREM. *Every field in which* -1 *is not a sum of squares can be ordered.*

PROOF. As in Example 2.3.2, use the Compactness Theorem. We must include in Σ the sentences reflecting the conditions (2.4). The assumption of the theorem guarantees that every finite subset of Σ has a model.

2.4. The Countable Axiom of Choice

Very often, especially when dealing with real numbers, one does not need the full Axiom of Choice but only its weaker variant, the Countable Axiom of Choice.

COUNTABLE AXIOM OF CHOICE. *Every countable family of nonempty sets has a choice function.*

We shall give several examples of applications of this principle.

2.4.1. *Example: The Countable Axiom of Choice implies that every infinite set has a countable subset.*

PROOF. Let S be an infinite set. Consider all finite one-to-one finite sequences

$$\langle a_0, a_1, \ldots, a_k \rangle$$

of elements of S. The Countable Axiom of Choice picks out one k-sequence for each natural number; more exactly:

$$\mathscr{F} = \{A_k : k \in \omega\},$$

where

$$A_k = \{\langle a_0, \ldots, a_k \rangle : a_0, \ldots, a_k \text{ distinct elements of } S\},$$

and $\mathscr{F}$ has a choice function: $f(A_k) \in A_k$ for all k. The union of all the chosen finite sequences is obviously countable.

2.4.2. *Example: The Countable Axiom of Choice implies that the union of countably many countable sets is countable.*

PROOF. See Problem 1.6.

COROLLARY 1. *The set of all real numbers is not a countable union of countable sets.*

COROLLARY 2. *The first uncountable ordinal ω_1 is not a limit of a countable increasing sequence of ordinals.*

(Both things can happen in the absence of the Axiom of Choice, as we shall see later; cf. Theorem 10.6.)

2.4.3. *Example: Topological properties of the real line.*

There are two kinds of definitions of the basic topological properties of the real line: (1) $\varepsilon-\delta$-definitions and (2) definitions using limits of sequences.

(a). *Closed sets*: A point x is in the closure of a set A if

(1) every neighborhood of x intersects A;

(2) $x = \lim_{n\to\infty} x_n$ for some sequence $\{x_n\}$ of points in A.

(b). *Continuous functions*: A function f is continuous at a point x if

(1) $\forall\varepsilon\, \exists\delta$ etc.

(2) whenever $\lim x_n = x$, then $\lim f(x_n) = f(x)$.

(c). *Compact sets*: A set A is compact if

(1) A is closed and bounded;

(2) every sequence $\{x_n\}$ of points in A has a convergent subsequence with $\lim x_n \in A$.

(Notice that the Heine–Borel theorem can be proved for the closed and bounded sets – see Problem 25).

PROPOSITION. *In the presence of the Countable Axiom of Choice, the definitions under* (1) *and* (2) *are equivalent in each of the cases* (a), (b), (c).

PROOF. One direction is trivial in each case.

(a). If x is in the closure of A, then each $(1/n)$-neighborhood of x intersects A. We can choose x_n by the Countable Axiom of Choice.

(b). If f is discontinuous at x, then there is $\varepsilon > 0$ such that for each n, we can choose x_n in the $(1/n)$-neighborhood of x such that $|f(x_n)-f(x)| \geqslant \varepsilon$. Then $f(x) = \lim f(x_n)$ is false.

(c). Similar.

2.4.4. *Example: The Countable Axiom of Choice implies that every subspace of a separable metric space is separable.*

PROOF. A metric space S is separable if it has a countable dense subset $\{x_0, x_1, \ldots, x_n, \ldots\}$. The family

$$\{U_{mn} : m \in \omega, n \in \omega\},$$

where U_{mn} is the $(1/m)$-neighborhood of x_n, is a countable base of S. If T is a subspace of S, then the family

$$\{U_{mn} \cap T : m \in \omega, n \in \omega\}$$

is a countable base of T. By the Countable Axiom of Choice, we can choose $y_{mn} \in U_{mn} \cap T$ for each m and n, thus getting a countable dense subset of T.

2.4.5. *Example: The Countable Axiom of Choice implies the countable additivity of the Lebesgue measure and of meager sets.*

PROOF. For meager sets, see Problem 1.7. As for the Lebesgue measure, suffice it to say that in the standard development of the theory of Lebesgue measurability, the only applications of the Axiom of Choice have the following form: If $\mathscr{F}_i$, $i \in \omega$, are nonempty families of open sets, then there is a sequence of open sets $\{G_i\}$ with $G_i \in \mathscr{F}_i$ for each i.

2.4.6. *Example: Borel sets*

There are two definitions of Borel sets of real numbers:

(a). The family B of Borel sets is the smallest countably additive algebra of sets of reals containing the open sets.

Among the standard properties of Borel sets are:

(i). $\mathscr{B} = \bigcup_{\alpha<\omega_1} \mathscr{B}_\alpha$, where $\mathscr{B}_0$ are the open sets and $\mathscr{B}_\alpha$ are all countable unions of elements of $\bigcup_{\gamma<\alpha} \mathscr{B}_\gamma$ and their complements.

(ii). $\mathscr{B}_\alpha \subsetneqq \mathscr{B}_{\alpha+1}$, for each $\alpha < \omega_1$.

If the Axiom of Choice is absent, then we have to replace (i) by a weaker property, with ω_1 replaced by some unspecified ordinal (the least ordinal ϑ such that $B_{\vartheta+1} = B_\vartheta$). The property (ii) cannot be proved either (see the remark at the end of Example 2.4.2.).

(b). In recursion theory, one defines the hierarchy of Borel sets (sets hyperarithmetical in some real) using codes. Roughly speaking, each Borel set has a code, but without choice, one cannot choose a code for each Borel set. Thus one cannot prove that the family of Borel sets is closed under arbitrary countable unions.

In the presence of the Countable Axiom of Choice, both definitions of Borel sets are equivalent, and the properties (i) and (ii) can be proved.

2.4.7. *Principle of Dependent Choices*

Finally, let us mention a principle closely related to the Countable Axiom of Choice. Its meaning is that one is allowed to make a countable number of consecutive choices.

PRINCIPLE OF DEPENDENT CHOICES. *If ρ is a relation on a nonempty set A such that for every $x \in A$ there exists $y \in A$ with $x\rho y$, then there is a sequence $\{x_n\}$ of elements of A such that*

$$x_0 \rho x_1, \quad x_1 \rho x_2, \ldots, x_n \rho x_{n+1}, \ldots$$

PROPOSITION. *The Principle of Dependent Choices implies the Countable Axiom of Choice.*

PROOF. Let $S_0, S_1, \ldots, S_n, \ldots$ be a countable family of nonempty sets. To find a choice function, let A be the set of all finite sequences

$$s = \langle x_0, x_1, \ldots, x_k \rangle$$

such that $x_0 \in S_0, \ldots, x_k \in S_k$, and let $s \rho t$ just in case $s = \langle x_0, \ldots, x_k \rangle$ and $t = \langle x_0, \ldots, x_{k+1} \rangle$. An application of the Principle of Dependent Choices gives a choice function for the family $\{S_n\}$.

As an application of the Principle of Dependent Choices, we give a characterization of well-orderings:

PROPOSITION. *A linear ordering $<$ of a set P is a well-ordering if and only if P has no infinite descending sequence*

$$x_0 > x_1 > x_2 > \ldots > x_n > \ldots .$$

PROOF. The condition is obviously necessary.

To show that it is sufficient, assume that $<$ is not a well-ordering and use the Principle of Dependent Choices to construct the descending sequence.

2.5. Cardinal numbers

Sets can be compared by their cardinality, using the definition

$|X| \leqslant |Y|$ iff there exists a one-to-one mapping of X *into* Y,

$|X| = |Y|$ iff there exists a one-to-one mapping of X *onto* Y.

We may run into difficulties when we want to define the symbol $|X|$, but that we shall discuss later.

First, we prove that the relation $\leqslant$ is a partial ordering:

THEOREM 2.3. *If $|X| \leqslant |Y|$ and $|Y| \leqslant |X|$, then $|X| = |Y|$.*

PROOF. It suffices to prove that if $A_1 \subseteq B \subseteq A$ and $|A_1| = |A|$, then $|B| = |A|$. Let f be a one-to-one function of A onto A_1; let

$$A_0 = A, \quad A_1 = f''A_0, \quad A_2 = f''A_1, \ldots,$$
$$B_0 = B, \quad B_1 = f''B_0, \quad B_2 = f''B_1, \ldots .$$

If we let

$$g(x) = \begin{cases} f(x) & \text{if } x \in A_n - B_n \text{ for some } n, \\ x & \text{otherwise,} \end{cases}$$

then g is a one-to-one mapping of A onto B.

There is not much more one can prove about the ordering $\leqslant$. The well-orderable sets are equivalent to ordinal numbers and so we can define *cardinals* of well-ordered sets, using initial ordinals (i.e. those which are not equivalent to smaller ordinals). As is customary, $\aleph_\alpha = \omega_\alpha$ is the αth infinite cardinal number. The elementary arithmetic of alephs is fairly simple, due to the following formula:

THEOREM 2.4. *For all* α, $\aleph_\alpha \cdot \aleph_\alpha = \aleph_\alpha$.

PROOF. There is a canonical one-to-one mapping of ω_α onto $\omega_\alpha \times \omega_\alpha$. We define an ordering $\prec$ of $\omega_\alpha \times \omega_\alpha$ by:

$$\begin{aligned}(\alpha_1, \alpha_2) \prec (\beta_1, \beta_2) \text{ iff } & \max(\alpha_1, \alpha_2) < \max(\beta_1, \beta_2) \\ & \vee [\max(\alpha_1, \alpha_2) = \max(\beta_1, \beta_2) \wedge \alpha_1 < \beta_1] \\ & \vee [\max(\alpha_1, \alpha_2) = \max(\beta_1, \beta_2) \wedge \alpha_1 = \beta_1 \wedge \alpha_2 < \beta_2].\end{aligned}$$

The relation $\prec$ well-orders $\omega_\alpha \times \omega_\alpha$ and a proof by transfinite induction shows that the order-type of this ordering is ω_α.

As a consequence, we have

$$\aleph_\alpha + \aleph_\beta = \aleph_\alpha \cdot \aleph_\beta = \max(\aleph_\alpha, \aleph_\beta),$$
$$\aleph_\alpha + \aleph_\alpha = \aleph_\alpha \cdot \aleph_\alpha = \aleph_\alpha.$$

This means that in the presence of the Axiom of Choice, we have

(2.5) $$\mathfrak{m} + \mathfrak{m} = \mathfrak{m} \cdot \mathfrak{m} = \mathfrak{m},$$

for every infinite cardinal $\mathfrak{m}$. There will be more discussion about (2.5) in Chapter 11.

One of the properties in whose proof one needs the Axiom of Choice is *regularity* of successor alephs $\aleph_{\alpha+1}$. (An infinite cardinal κ is *regular* if $\operatorname{cf} \kappa = \kappa$, where $\operatorname{cf} \kappa$, the cofinality of κ, is the least ordinal α such that κ is a limit of an increasing α-sequence $\{\gamma_\xi : \xi < \alpha\}$ of ordinals less than κ. If $\operatorname{cf} \kappa < \kappa$ then κ is *singular*.) E.g., to show that $\aleph_1$ is regular, one uses the fact that the countable union of countable sets is countable; see Example 2.4.2. Similarly for $\aleph_{\alpha+1}$.

Another problem involving the Axiom of Choice is the definition of infinite sums and products of cardinal numbers. To have a meaningful definition of, say,

$$m_1 \cdot m_2 \cdot \ldots \cdot m_n \cdot \ldots \quad (n \in \omega),$$

we have to know that

$$|\mathsf{X}\{M_n : n \in \omega\}| = |\mathsf{X}\{M'_n : n \in \omega\}| \tag{2.6}$$

whenever $|M_n| = |M'_n|$ for each n. The proof of (2.6), however, requires the Axiom of Choice. For a counterexample, assume that $|M'_n| = 2$ and that $\{M'_n\}$ does not have a choice function; compare this with $M_n = \{0, 1\}$ for each n.

To complete this section, we will look into the questions of finiteness. While by the standard definition a set is finite if it is equivalent to a natural number, there is another definition of finiteness, which is equivalent to the standard definition if the Axiom of Choice is true:

DEFINITION 2.6. A set X is *Dedekind finite* if there is no one-to-one mapping of X onto a proper subset of X.

How this notion is related to finiteness is clear from the following characterization of Dedekind finite sets:

LEMMA 2.7. *X is Dedekind finite if and only if X does not have a countable subset.*

PROOF. If X contains a countable subset then obviously X is Dedekind infinite.

On the other hand, if X is Dedekind infinite, then let f be a mapping of X onto a proper subset of X and $a \in X$ which is not in the range of f. The set $\{a, f(a), f(f(a)), \ldots\}$ is a countable subset of X.

2.6. Problems

1. The following statement is equivalent to the Axiom of Choice: The Cartesian product of any family of nonempty sets is nonempty.

2. Prove the following self-refinement of Zorn's Lemma: If every chain in $(P, <)$ has an upper bound, then there is a maximal element above any given element of P.

3. Prove the following self-refinement of Tukey's Lemma: If $\mathscr{F}$ has finite character, then every $X \in \mathscr{F}$ is included in a maximal $Y \in \mathscr{F}$.

4. The following statement is equivalent to the Axiom of Choice: Every family of sets contains a maximal subfamily consisting of mutually disjoint sets.

[*Hint*: Let S be a family of disjoint sets; find a choice function on S. This can be obtained using a maximal disjoint subfamily of $\mathscr{F} = \{\{(0, a), (1, A)\} : A \in S \text{ and } a \in A\}$.]

Let us call a set X of real numbers *independent* if no linear combination $r_1 a_1 + \ldots + r_k a_k$ with distinct $a_1, \ldots, a_k \in X$ and nonzero rational coefficients $r_1, \ldots, r_k$ is equal to zero. A set H of real numbers is a *Hamel basis* if every real number x can be uniquely written in the form $x = r_1 h_1 + \ldots + r_k h_k$, where $h_1, \ldots, h_k \in H$ and $r_1, \ldots, r_k$ are rational numbers.

5. Using some form of the Axiom of Choice, prove that a Hamel basis exists.

6. There is a discontinuous function from R to R which satisfies the equation $f(x+y) = f(x)+f(y)$.

[*Hint*: Use Problem 5: Let $h \in H$ and for each $x = rh + r_1 h_1 + \ldots + r_k h_k$, let $f(x) = r$.]

7. Using the Axiom of Choice, prove the following: If B_1 and B_2 are both bases of a vector space V, then $|B_1| = |B_2|$.

[*Hint*: There is a mapping f of B_1 into the finite subsets of B_2 and every b_2 is in some $f(b_1)$; and vice versa.]

8. Tychonoff's Theorem implies the Axiom of Choice.

[*Hint*: Find a choice function on a given family $\{X_i : i \in I\}$. Adjoin a fixed element a to each X_i and topologize $Y_i = X_i \cup \{a\}$ by letting only Y_i, X_i and finite sets be closed. In the product of Y_i's, consider, for each i, the closed set Z_i of all f such that $f(i) \in X_i$. The family $\{Z_i : i \in I\}$ is a filter-base and by Tychonoff's Theorem has a nonempty intersection.]

9. The following statement implies the Axiom of Choice: The product of any number of copies of the same compact space is compact.

[*Hint*: Find a choice function on a disjoint family $\{A_i : i \in I\}$. Let S be the union of all A_i with the least topology such that all A_i are closed. In the product $\times\{S : i \in I\}$, consider, for each i, the closed set Z_i of all f such that $f(i) \in A_i$.]

10. The following statement is equivalent to the Axiom of Choice: Every lattice with a unit element and at least one other element contains a maximal ideal.

[*Hint*: Use this to prove Tukey's Lemma. Let $\mathscr{F} \subseteq \mathscr{P}(A)$ have finite character. Let $L = \mathscr{F} \cup \{A\}$ and let $X \wedge Y = X \cap Y$, $X \vee Y =$ either $X \cup Y$ or A (if $X \cup Y \notin \mathscr{F}$). A maximal ideal in the lattice L gives a maximal element of $\mathscr{F}$.]

11. The following statement is equivalent to the Axiom of Choice: For every set S of nonzero elements of a Boolean algebra B, there is a maximal ideal disjoint from S.

[*Hint*: Use this to prove Tukey's lemma. Let $\mathscr{F} \subseteq \mathscr{P}(A)$ have finite character. Put $B =$ the set algebra $\mathscr{P}(A)$, $\mathscr{S} = \mathscr{P}(A) - \mathscr{F}$. A maximal ideal disjoint from $\mathscr{S}$ gives a maximal element of $\mathscr{F}$.]

12. The Stone Representation Theorem implies the Prime Ideal Theorem.

[*Hint*: Let B be a Boolean algebra, $\pi''B = \mathscr{F}$ an isomorphic set algebra. Let $p \in \bigcup \mathscr{F}$ and let $U = \{u \in B : p \in \pi(u)\}$.]

The next five problems establish the equivalence of the Prime Ideal Theorem with weaker versions of Tychonoff's Theorem.

13. If U is an ultrafilter over a Hausdorff space, then the intersection $\bigcap \{\overline{X} : X \in U\}$ contains at most one point.

14. The Ultrafilter Theorem implies that every product of compact Hausdorff spaces is nonempty.

[*Hint*: Let $S = \mathsf{X}\{S_i : i \in I\}$. Let Z be the set of all functions with $\operatorname{dom}(f) \subseteq I$ and $f(i) \in S_i$; $Z_i = \{f \in Z : i \in \operatorname{dom}(f)\}$. Let $\mathscr{F}$ be the filter over Z generated by Z_i, $i \in I$, and let $U \supseteq \mathscr{F}$ be an ultrafilter. Let U_i be the projection of U onto S_i, each U_i is an ultrafilter over S_i. For each i, the intersection $\bigcap \{\overline{X} : X \in U_i\}$ contains exactly one point $x_i \in S_i$.]

15. The Ultrafilter Theorem implies that every product of compact Hausdorff spaces is compact.

[*Hint*: The product is nonempty by Problem 14. In the proof of Tychonoff's Theorem in Section 2.2, the first part uses only the Ultrafilter Theorem. The use of Axiom of Choice in point 2 is eliminated by Problem 13.]

16. The following statement implies the Prime Ideal Theorem: The product of any family of discrete two-point spaces is compact.

[*Hint*: Let B be a Boolean algebra, let $S = \mathsf{X}\{\{u, -u\} : u \in B\}$. If A is a finite subalgebra of B and I a prime ideal on A, let $X_I = \{x \in S : x(u) \in I$ for each $u \in A\}$; further let $X_A = \bigcup \{X_I : I$ a prime ideal on $A\}$. The family $\{X_A : A$ a finite subalgebra$\}$ is a filter-base of closed sets; its intersection gives a prime ideal of B.]

17. The following statement implies the Prime Ideal Theorem: The generalized Cantor space $\{0, 1\}^I$ is compact for any I.

[*Hint*: Use this to prove the statement from Problem 16. It suffices to show that every product of two point sets is nonempty. Let S be a set of disjoint (unordered) pairs $p = \{a, b\}$ and let $I = \bigcup \{p : p \in S\}$. In the product $\{0, 1\}^I$, the sets $X_p = \{f : f(a) \neq f(b) \text{ if } \{a, b\} = p\}$ are closed and form a filter-base. The intersection gives a choice function on S.]

*18. The Prime Ideal Theorem implies the Stone–Čech Compactification Theorem.

[*Hint*: The space βX is the closure of the natural embedding of X into the product $[0, 1]^{F(X)}$, where $F(X)$ is the set of all continuous functions of X into $[0, 1]$. The product is compact by the Prime Ideal Theorem.]

A (real-valued) *measure* on a Boolean algebra B is a nonnegative function μ on B such that $\mu(0) = 0$, $\mu(1) = 1$, and $\mu(a+b) = \mu(a)+\mu(b)$ whenever $a \cdot b = 0$.

**19. The Hahn–Banach Theorem follows from the Prime Ideal Theorem. Moreover, it is equivalent to the statement that every Boolean algebra admits a real-valued measure.

In the following five problems, C_n is the following statement:

Every family of n-element sets has a choice function.

An *n-coloring* of a graph is a partition of its vertices into n classes such that no two vertices in one class are joined by an edge. P_n is the following statement:

For every graph G, if every finite subgraph is n-colorable, then G is n-colorable.

20. The Prime Ideal Theorem implies P_n, for any n.

[*Hint*: Use the Compactness Theorem.]

21. P_{n+1} implies P_n.

[*Hint*: Let G be a graph satisfying the hypothesis of P_n. Add a new vertex and join it to all vertices of G; now use P_{n+1}.]

22. P_n implies C_n.

[*Hint*: Let S be a family of disjoint n-element sets. Make its union into a graph by joining those and only those vertices which are in the same set in S. Use P_n to get a choice function.]

23. C_2 implies P_2.

[*Hint*: Let G be a graph satisfying the hypothesis of P_2. G does not contain any cycles of odd length. Thus every component of G is 2-colorable. Use C_2 to choose a 2-coloring for each component.]

**24. P_3 implies the Prime Ideal Theorem.

25. The Heine–Borel Theorem can be proved without using the Axiom of Choice: Every descending sequence of closed subsets of [0, 1] has a non-empty intersection.

*26. Prove Urysohn's Lemma, using the Principle of Dependent Choices: If A, B are disjoint closed sets in a T_4-space S, then there is a continuous function from S to [0, 1] which takes the value 1 everywhere in A and 0 everywhere in B.

[*Hint*: Follow the standard proof in a topology textbook.]

*27. Use the Countable Axiom of Choice to prove the following theorem: The set R of all real numbers is not a union of countably many subsets, S_n, $n \in \omega$, such that $|S_n| < |R|$ for each n.

[*Hint*: $|R| = 2^{\aleph_0} = |R|^{\aleph_0}$. Let ${}^{\omega}R$ be the set of all functions from ω into R; let $S_n \subset {}^{\omega}R$, $|S_n| < |R|$, for all $n \in \omega$. Let $T_n \subseteq R$ be the set $\{f(n) : f \in S_n\}$. Pick $g(n) \in R - T_n$, for each n; then $g \notin S_n$ for any n.]

2.7. Historical remarks

The Well-ordering Theorem was proved by Zermelo [1904]. The Maximal Principle I is due to Kuratowski [1922] and Zorn [1935]. It is known to the public as Zorn's Lemma (cf. e.g., Frampton [1970]). Similarly, the Maximal Principle II was formulated independently by Teichmüller [1939] and Tukey [1940].

A large collection of equivalents of the Axiom of Choice was accumulated by Rubin and Rubin [1963]. The proof of Tychonoff's Theorem given in this book is due to Bourbaki. The Nielsen–Schreier Theorem was proved by Nielsen for finitely generated free groups, and by Schreier in the general case.

Equivalents of the Prime Ideal Theorem were considered by several people; let us mention at least Tarski [1954a], Scott [1954b], Rado [1949], and Henkin [1954] – the latter proved its equivalence to the Compactness Theorem. The Stone Representation Theorem and its equivalence to the

Prime Ideal Theorem is due to Stone [1936]. The Order Extension Principle was proved by Marczewski [1930]. The observation that in the proof the Axiom of Choice can be replaced by the Prime Ideal Theorem is due to Scott; a similar observation in case of the Artin–Schreier Theorem is due to Tarski [1954a].

The Principle of Dependent Choices was formulated by Bernays [1942].

The example in Problem 4 was given by Vaught [1952]. The proof that Tychonoff's Theorem, or its weaker version in Problem 9, implies the Axiom of Choice, is due to Kelley [1950] and Ward [1962], respectively. The example in Problem 10 was given by Scott [1954a]; the statement in Problem 11 was considered by Mrówka [1955], Rubin and Rubin [1963], and Rousseau [1965]. The equivalence of the Prime Ideal Theorem to Tychonoff's Theorem for compact Hausdorff spaces is due to Łoś and Ryll-Nardzewski [1955], and Rubin and Scott [1954]. The stronger version of this equivalence in Problem 17 is due to Mycielski [1964a]. The proof of the Hahn–Banach Theorem from the Prime Ideal Theorem was given by Łoś and Ryll-Nardzewski [1951] and Luxemburg [1962]; the equivalence stated in Problem 19 is due to Ryll–Nardzewski (unpublished) and Luxemburg [1969]. The statements P_n were considered by Mycielski [1961] and Levy [1963a]; the theorem in Problem 24 is due to Läuchli [1971].

CHAPTER 3

CONSISTENCY OF THE AXIOM OF CHOICE

3.1. Axiomatic systems and consistency

The discussion in Chapter 2 makes it obvious that the benefits of the Axiom of Choice outweigh the inconveniences; one can easily put up with nonmeasurable sets and their likes in order to have another device to use in mathematical proofs. It would be naturally nice if one could prove the Axiom of Choice, from other, less controversial principles. It was realised very soon that this is very unlikely, since the Axiom of Choice is so different.

The next best thing to proving the Axiom of Choice is to show that it is not contradictory. That is to show that it is *consistent* with other principles of mathematics.

In general, the problem of consistency of a certain statement σ consists in showing that σ cannot be disproved in a given axiomatic system (provided the system itself is not contradictory). The usual way is to give an *interpretation* of the *language* of the system, in which all the axioms are true, together with the statement σ.

3.2. Axiomatic set theory

The language of set theory consists of two primitive symbols, $=$ and $\in$, and predicates, operations and constants defined from the primitive symbols. The formulas are built by means of logical connectives and quantifiers.

Technically, all objects occuring in axiomatic set theory are sets. However, we shall sometimes informally use *classes*, which represent collections of sets of the form

$$C = \{x : \phi(x, p)\}$$

(where p is a parameter).

In the customary way, we define the standard set-theoretical operations, unordered and ordered pairs, the latter by the definition

$$(u, v) = \{\{u\}, \{u, v\}\},$$

relations and functions. Let us mention that the symbols

$$\operatorname{dom}(f), \qquad \operatorname{rng}(f), \qquad f''X, \qquad f|X, \qquad {}^{X}Y, \qquad \mathscr{P}(X), \qquad 0$$

denote respectively the domain of f, the range of f, the image of X under f $(= \{f(x) : x \in X\})$, the restriction of f to X, the set of all functions from X to Y, the power set of X $(= \{u : u \subseteq X\})$ and the empty set.

The axioms of Zermelo–Fraenkel axiomatic set theory ZF are as follows:

A1. EXTENSIONALITY:

$$\forall u\,(u \in X \leftrightarrow u \in Y) \rightarrow X = Y.$$

A2. PAIRING:

$$\forall u\,\forall v\,\exists x\,\forall z\,(z \in X \leftrightarrow z = u \vee z = v).$$

A3. COMPREHENSION:

$$\forall \vec{p}\,\forall X\,\exists Y\,\forall u\,(u \in Y \leftrightarrow u \in X \wedge \phi(u, \vec{p})),$$

for any formula of set theory.[1]

A4. UNION:

$$\forall x\,\exists y\,\forall z\,\forall u\,(u \in z \wedge z \in X \rightarrow u \in Y).$$

A5. POWER-SET:

$$\forall X\,\exists Y\,\forall u\,(u \subseteq X \rightarrow u \in Y).$$

A6. REPLACEMENT: Let ϕ be a formula such that $\mathscr{F} = \{(x, y) : \phi(x, y, \vec{p})\}$ is a function. Then

$$\forall X\,\exists Y\,(\mathscr{F}''X \subseteq Y).$$

A7. INFINITY:

$$\exists Y\,[0 \in Y \wedge \forall u\,(u \in Y \rightarrow \{u\} \in Y)$$

A8. REGULARITY:

$$(\forall S \neq 0)\,(\exists x \in S)\,[x \cap S = 0]$$

[1] $\vec{p}$ stands for $p_1, \ldots, p_n$.

A set S is *transitive* if

$$\forall x\,(x \in S \to x \subseteq S). \tag{3.1}$$

(Similarly, a *transitive class* is a class which satisfies (3.1).) One can define *ordinal numbers* as transitive sets which are well-ordered. Then ordinal numbers are representatives of well-ordered sets and each ordinal α is the set of all smaller ordinals:

$$\alpha = \{\beta : \beta < \alpha\}.$$

It follows from the axiom of regularity that every set has a *rank*, and that the universe V is the union of sets

$$V = \bigcup_{\alpha \in \mathrm{On}} V_\alpha$$

where On denotes the class of all ordinals and V_α is the set of all sets of rank less than α:

$$V_0 = 0, \qquad V_{\alpha+1} = \mathscr{P}(V_\alpha), \qquad V_\alpha = \bigcup_{\beta<\alpha} V_\beta \; (\alpha \text{ limit}).$$

Transitive sets and classes are in a way a generalization of well-ordered sets. They have two important properties:

1. No two transitive sets are isomorphic (that is, there is no one-to-one mapping π such that $x \in y \leftrightarrow \pi x \in \pi y$). The only automorphism of a transitive set is the identity.

2. We can carry out induction and recursion on elements of a transitive class, like on ordinals (so-called $\in$-*induction* and $\in$-*recursion*). E.g., if

$$[(\forall x \in y)\, \phi(x)] \to \phi(y),$$

for every y in a transitive class T, then every $y \in T$ has the property ϕ (see Problem 3).

3.3. Transitive models of ZF

A natural way of interpretation of set theory in set theory is by the construction of *models*. We consider a class $\mathscr{M}$ and a binary relation E on $\mathscr{M}$, and interpret the formulas of set theory as follows: We define

$$\mathscr{M} \vDash \phi$$

($\mathscr{M}$ *satisfies* ϕ) inductively on the complexity of ϕ:

$\mathscr{M} \vDash x \in y$ iff $x\,E\,y$

$\mathscr{M} \vDash \forall x \in \phi$ iff for all $x \in \mathscr{M}$, $\mathscr{M} \vDash \phi(x)$,

etc.

Let us restrict ourselves to the simple case when E is the ordinary $\in$. For the purpose of getting models of set theory, it suffices to consider transitive classes only, in view of the following theorem, which is proved by transfinite or $\in$-recursion.

THEOREM 3.1. (Isomorphism Theorem). *If* $\mathcal{M} \models$ Extensionality, *then* $\mathcal{M}$ *is isomorphic to a transitive class.*

It is easy to see that if $\phi(\vec{x})$ is a formula without quantifiers, then it is true in a transitive model $\mathcal{M}$ if and only if ϕ is true in the universe. The same is true even for a formula ϕ with quantifiers, provided the only occurrence of quantifiers are restricted quantifications,

$$(\forall x \in y), \qquad (\exists x \in y).$$

Call a formula ϕ *absolute* (for a given transitive $\mathcal{M}$) if for all $x \in \mathcal{M}$,

$$\mathcal{M} \models \phi(x) \quad \text{iff} \quad \phi(x),$$

Similarly, an operation $\mathcal{F}$ is *absolute* if for all $x \in \mathcal{M}$,

$$\mathcal{F}(x) = \mathcal{F}^{\mathcal{M}}(x);$$

here

$$\mathcal{F}(x) = \{y : \phi(x, y)\},$$
$$\mathcal{F}^{\mathcal{M}}(x) = \{y \in \mathcal{M} : \mathcal{M} \models \phi(x, y)\},$$

where ϕ is the formula which defines $\mathcal{F}$.

We can use the above observation about restricted quantification and similar rules like 'if ϕ and $\mathcal{F}$ are absolute then $(\exists y \in \mathcal{F}(x))\, \phi$ is absolute' to verify that many fundamental properties are absolute. In particular, if $\mathcal{M}$ is a transitive class which is closed under the operation $\{\ \}$, i.e., $x, y \in \mathcal{M}$ implies $\{x, y\} \in \mathcal{M}$, then the following formulas are absolute:

> $x \subseteq y$, $x = 0$, x is a pair, x is a singleton, x is a binary relation, x is a function, x is transitive, x is an ordering, $\bigcup x$, $x - y$, $x \times y$, dom (x), rng (x), $\in \cap\, x^2$ $(= \{(u, v) : u \in x, v \in x, u \in v\})$, $x \cup y$, x is an ordinal,[2] x is a limit ordinal, x is a successor ordinal, etc.,

On the other hand, $\mathcal{P}(x)$ is not absolute, and neither is the formula 'x is a cardinal number'.

We say that a transitive class $\mathcal{M}$ is a *model of* ZF if it satisfies all the axioms of ZF. There is a useful criterion for transitive classes to be models

[2] Caution: we have to use the equivalence 'x is an ordinal iff x is transitive and linearly ordered by $\in$', which follows from the axiom of regularity.

of ZF. A transitive class $\mathcal{M}$ is *almost universal* if every subset of $\mathcal{M}$ is included in an element of $\mathcal{M}$:

$$(\forall S \subset \mathcal{M})\,(\exists Y \in \mathcal{M})\,[S \subseteq Y].$$

If $\mathcal{M}$ is transitive and almost universal, then $\mathcal{M}$ satisfies all the axioms of ZF except possibly Infinity and Comprehension. Infinity is obviously no problem since $\mathcal{M}$ contains infinite sets, and as soon as $\mathcal{M}$ satisfies Comprehension, it must satisfy Infinity also.

To verify that $\mathcal{M}$ satisfies Comprehension, it suffices to verify that it satisfies eight particular instances of it. The following operations are called Gödel operations (note that they are absolute for transitive classes closed under $\{\ \}$):

$$\begin{aligned}
\mathcal{F}_1(X, Y) &= \{X, Y\},\\
\mathcal{F}_2(X, Y) &= X - Y,\\
\mathcal{F}_3(X, Y) &= X \times Y,\\
\mathcal{F}_4(X) &= \operatorname{dom}(X),\\
\mathcal{F}_5(X) &= \in \cap X^2,\\
\mathcal{F}_6(X) &= \{(a, b, c) : (b, c, a) \in X\},\\
\mathcal{F}_7(X) &= \{(a, b, c) : (c, b, a) \in X\},\\
\mathcal{F}_8(X) &= \{(a, b, c) : (a, c, b) \in X\}.
\end{aligned}$$

THEOREM 3.2. *If $\mathcal{M}$ is transitive, almost universal and closed under Gödel operations then $\mathcal{M}$ is a model of* ZF.

PROOF. Let $\mathcal{M}$ be a transitive class, almost universal and closed under $\mathcal{F}_1, \ldots, \mathcal{F}_8$. We will show that $\mathcal{M}$ satisfies the axioms A1–A8.

Extensionality: $\mathcal{M}$ is transitive.

Pairing: $\mathcal{M}$ is closed under $\{\ \}$.

Union: If $X \in \mathcal{M}$, then $\bigcup^{\mathcal{M}} X = \bigcup X \subseteq \mathcal{M}$ and is a set; since $\mathcal{M}$ is almost universal, $\bigcup^{\mathcal{M}} X$ is included in some $Y \in \mathcal{M}$.

Power-set: The same argument, except that $\mathcal{P}^{\mathcal{M}}(X) = \mathcal{P}(X) \cap \mathcal{M}$.

Regularity: $\mathcal{M}$ is transitive and Regularity holds in the universe.

Infinity: There is a subset Y of $\mathcal{M}$ satisfying A7. Since $\mathcal{M}$ is almost universal, Y is included in some $X \in \mathcal{M}$. As soon as we have Comprehension, Y becomes a set in $\mathcal{M}$ satisfying A7.

Replacement: Let

$$\mathcal{F} = \{(x, y) \in \mathcal{M}^2 : \mathcal{M} \models \phi(x, y)\},$$

$$\mathcal{M} \models \mathcal{F} \text{ is a function.}$$

We show that $\mathcal{M} \models \forall X \exists Y (\mathcal{F}''X \subseteq Y)$. Let $\mathcal{G} : \mathcal{M} \to \mathcal{M}$ be the function defined by

$$y = \mathcal{G}(x) \quad \text{iff} \quad \mathcal{M} \models \phi(x, y).$$

If $X \in \mathcal{M}$, then there exists $Y \in \mathcal{M}$ such that $\mathcal{G}''X \subseteq Y$. Then $\mathcal{M} \models \mathcal{F}''X \subseteq Y$.

Comprehension: Let ϕ be a formula; we show that for every $X \in \mathcal{M}$, the set[3]

$$Y = \{(u_1, ..., u_n) \in X^n : \mathcal{M} \models \phi(\vec{u})\} \tag{3.2}$$

is in $\mathcal{M}$. We do that by showing that there is an operation $\mathcal{F}$, a composition of Gödel operations, such that for every $X \in \mathcal{M}$, if Y is as in (3.2), then $Y = \mathcal{F}(X, X_1, ..., X_k)$ for some $X_1, ..., X_k \in \mathcal{M}$. The proof is by induction on the complexity of ϕ.

(i). ϕ is an atomic formula. We will handle one typical case, the other cases being similar (including the cases involving a parameter). Let $\phi(u_1, ..., u_n)$ be $u_i \in u_j$, where $i \neq j$. Thus we have

$$Y = \{(u_1, ..., u_n) : u_1, ..., u_n \in X,\ u_i \in u_j\}.$$

We find $\mathcal{F}$ by induction on n. Let $v = (u_1, ..., u_{n-2})$.

Case 1. $i = n-1, j = n$. We have

$$\begin{aligned} Y &= \{(v, u_{n-1}, u_n) : v \in X^{n-2},\ (u_{n-1}, u_n) \in X^2,\ u_{n-1} \in u_n\} \\ &= \{(v, u_{n-1}, u_n) : v \in X^{n-2},\ (u_{n-1}, u_n) \in \mathcal{F}_5(X)\} \\ &= \mathcal{F}_6(\mathcal{F}_5(X) \times X^{n-2}). \end{aligned}$$

Case 2. $i = n, j = n-1$. Here

$$\begin{aligned} Y &= \{(v, u_{n-1}, u_n) : v \in X^{n-2},\ (u_n, u_{n-1}) \in \mathcal{F}_5(X)\} \\ &= \mathcal{F}_7(\mathcal{F}_5(X) \times X^{n-2}). \end{aligned}$$

Case 3. $i, j \neq n$. By the induction hypothesis, there is $\mathcal{G}$ such that

$$\{(u_1, ..., u_{n-1}) : u_1, ..., u_{n-1} \in X,\ u_i \in u_j\} = \mathcal{G}(X).$$

So we have $Y = \mathcal{G}(X) \times X$.

Case 4. $i, j \neq n-1$. By the induction hypothesis, there is $\mathcal{G}$ such that

$$\{(v, u_n) : v \in X^{n-2}, u_n \in X,\ u_i \in u_j\} = \mathcal{G}(X).$$

Thus we have

$$Y = \{(v, u_{n-1}, u_n) \in X^n : u_i \in u_j\} = \mathcal{F}_8(\mathcal{G}(X) \times X).$$

[3] For simplicity, we drop the parameters in ϕ.

(ii). The connectives. Let $Y_i = \{(u_1, \ldots, u_n) \in X^n : \phi_i(\vec{u})\}$, $i = 1, 2$. Then

$$\{(u_1, \ldots, u_n) \in X^n : \neg\, \phi_1(\vec{u})\} = X^n - Y_1 = \mathscr{F}_2(X^n, Y_1),$$

$$\{(u_1, \ldots, u_n) \in X^n : \phi_1(\vec{u}), \phi_2(\vec{u})\} = Y_1 \cap Y_2 = \mathscr{F}_2(Y_1, \mathscr{F}_2(Y_1, Y_2)).$$

(iii). The existential quantifier. Consider a formula

$$\exists v\, \phi(u_1, \ldots, u_n, v)$$

and assume that there is $\mathscr{G}$ such that for every $Z \in \mathscr{M}$ there exist $Z_1, \ldots, Z_k \in \mathscr{M}$ such that

$$\{(\vec{u}, v) \in Z^{n+1} : \mathscr{M} \models \phi(\vec{u}, v)\} = \mathscr{G}(Z, Z_1, \ldots, Z_k).$$

We will show that for every $X \in \mathscr{M}$, there exist $X_1, \ldots, X_{k+1} \in \mathscr{M}$ such that

(3.3) $\{u \in X^n : \mathscr{M} \models \exists v\, \phi(\vec{u}, v)\} = X^n \cap \operatorname{dom}(\mathscr{G}(X_{k+1}, X_1, \ldots, X_k)).$

We need the following lemma:

LEMMA 3.3. *Let $\phi(x, \vec{y})$ be a formula. For each set S_0 there is $S \supseteq S_0$ such that for all $\vec{y} \in S$,*

$$\exists x\, \phi(x, \vec{y}) \to (\exists x \in S)\, \phi(x, \vec{y}).$$

PROOF. If C is a class, denote

$$\hat{C} = \{x \in C : (\forall y \in C)\, (\operatorname{rank}(x) \leqslant \operatorname{rank}(y))\};$$

then $\hat{C}$ is a set and $\hat{C} \subseteq C$. Let

$$C_y = \{x : \phi(x, y)\},$$

and let

$$\mathscr{T}(X) = X \cup \bigcup \{\hat{C}_y : y \in X\}.$$

By recursion, let $S_1 = \mathscr{T}(S_0), \ldots, S_{n+1} = \mathscr{T}(S_n), \ldots, S = \bigcup_{n=0}^{\infty} S_n$. The set S will do because if $\vec{y} \in S$ and $\exists x\, \phi(x, \vec{y})$ then $(\exists x \in S)\, \phi(x, \vec{y})$.

Now let $X \in \mathscr{M}$. Using Lemma 3.3, there exists $S \subseteq \mathscr{M}$ such that $S \supseteq X$ and for all $\vec{u} \in X$,

$$(\exists v \in \mathscr{M})\, [\mathscr{M} \models \phi(\vec{u}, v)] \quad \text{iff} \quad (\exists v \in S)\, [\mathscr{M} \models \phi(\vec{u}, v)].$$

Since $\mathscr{M}$ is almost universal, there exists $X_{k+1} \in \mathscr{M}$ such that $X_{k+1} \supseteq S$. Then for all $\vec{u} \in X$,

$$(\exists v \in \mathscr{M})\, [\mathscr{M} \models \phi(\vec{u}, v)] \quad \text{iff} \quad (\exists v \in X_{k+1})\, [\mathscr{M} \models \phi(\vec{u}, v)].$$

Now let $X_1, \ldots, X_k \in \mathcal{M}$ be such that $\mathcal{G}(X_{k+1}, X_1, \ldots, X_k) \subseteq X_{k+1}^{n+1}$ and for all $\vec{u}, v \in X_{k+1}$,

$$\mathcal{M} \models \phi(\vec{u}, v) \quad \text{iff} \quad (\vec{u}, v) \in \mathcal{G}(X_{k+1}, X_1, \ldots, X_k).$$

It follows that for all $\vec{u} \in X$,

$$\begin{aligned} \mathcal{M} \models \exists v\, \phi(\vec{u}, v) \quad &\text{iff} \quad (\exists v \in \mathcal{M})\, \mathcal{M} \models \phi(\vec{u}, v) \\ &\text{iff} \quad (\exists v \in X_{k+1})\, \mathcal{M} \models \phi(\vec{u}, v) \\ &\text{iff} \quad (\exists v \in X_{k+1})\, (\vec{u}, v) \in \mathcal{G}(X_{k+1}, X_1, \ldots, X_k) \\ &\text{iff} \quad (u_1, \ldots, u_n) \in X^n \cap \operatorname{dom}(\mathcal{G}(X_{k+1}, X_1, \ldots, X_k)), \end{aligned}$$

which proves (3.3), and completes the proof of the theorem.

3.4. The constructible universe

It will be proved in this section that there exists a least transitive model of ZF which contains all ordinals and that this least model satisfies the Axiom of Choice. That will establish the consistency of the Axiom of Choice with the axioms of ZF.

By the results of the preceding section, the ordinals are absolute, and every transitive model has to be closed under Gödel operations, which are absolute. This reasoning leads to the following definition: The *closure* $\operatorname{cl}(S)$ of a set S is the least $S' \supseteq S$ closed under Gödel operations; in fact,

$$\operatorname{cl}(S) = S_0 \cup S_1 \cup \ldots \cup S_n \cup \ldots, \qquad n \in \omega,$$

where

$$S_0 = S, \quad S_{n+1} = S_n \cup \{\mathcal{F}_i(x, y) : i = 1, \ldots, 8,\ x, y \in S_n\}.$$

Define

$$\begin{aligned} L_0 &= 0, \\ L_\alpha &= \bigcup_{\beta < \alpha} L_\beta \quad \text{if } \alpha \text{ is a limit ordinal,} \\ L_{\alpha+1} &= \mathcal{P}(L_\alpha) \cap \operatorname{cl}(L_\alpha \cup \{L_\alpha\}), \end{aligned} \tag{3.4}$$

and let L be the class

$$L = \bigcup_{\alpha \in \mathrm{On}} L_\alpha .$$

It is easy to see that L is transitive, almost universal and closed under Gödel operations; hence:

THEOREM 3.4. *L is a model of* ZF.

L is called the *constructible universe* and the elements of L are *constructible sets*.

The class L can obviously be well-ordered: one simply goes through the construction (3.4) and keeps enumerating the constructible sets by ordinals; this is possible because every new constructible set is either $\mathscr{F}_i(x, y)$, where x and y have been enumerated, or L_α. Consequently, if every set is constructible, then the universe can be well-ordered, and the Axiom of Choice holds.

AXIOM OF CONSTRUCTIBILITY: $V = L$, *i.e., every set is constructible.*

THEOREM 3.5. $L \models$ Axiom of Constructibility, *and therefore L is a model of* ZF+Axiom of Choice.

PROOF. Since every set in L is constructible, it suffices to show that the formula

$$x \text{ is constructible}$$

is absolute for L. This can be done in a succession of simple observations:

(a). The operations $\mathscr{G}_i(X) = \mathscr{F}_i''(X \times X)$, $i = 1, \ldots, 8$, are absolute.

(b). If $X \in L$, then $\mathscr{G}_i(X) \in L$, i $= 1, \ldots, 8$.

(c). $\omega \in L$.

(d). The following formula $\phi(g, x)$ is absolute:

$$\operatorname{dom}(g) = \omega \wedge g(0) = X \wedge (\forall n \in \omega)\, [g(n+1) = g(n) \cup \textstyle\bigcup_{i=1}^{8} \mathscr{G}_i(g(n))].$$

(e). If $X \in L$ and $\phi(g, x)$ then $g \in L$.

(f). The operation $\mathscr{H}(X) = \mathscr{P}(X) \cap \operatorname{cl}(X \cup \{X\})$ is absolute.

(g). If $X \in L$, then $\mathscr{H}(X) \in L$.

(h). The following formula $\psi(f, \alpha)$ is absolute,

$$\begin{aligned} \operatorname{dom} f = \alpha + 1 \wedge f(0) = 0 \\ &\wedge (\forall \beta \in \operatorname{dom} f)\, [\beta \text{ limit} \rightarrow f(\beta) = f''\beta] \\ &\wedge (\forall \beta < \alpha)\, [f(\beta+1) = \mathscr{H}(f(\beta))]. \end{aligned}$$

(i). If $\alpha \in L$ and $\psi(f, \alpha)$, then $f \in L$.

(j). The formula $X \in L_\alpha$ is absolute.

The proof of Theorem 3.5 actually gives a somewhat stronger result:

If $\mathscr{M}$ is a transitive model of ZF, *then the formula* 'x is constructible' *is absolute for $\mathscr{M}$.* (See Problem 10.)

Thus if $\mathscr{M}$ is a transitive model of ZF and contains all ordinals, then $\mathscr{M}$ contains all constructible sets. Hence L is the least transitive model of ZF containing all ordinals.

3.5. Problems

1. For every set S, there exists the *transitive closure* of S, $\mathrm{TC}(S)$, which is the least transitive $T \supseteq S$.

[*Hint*: $\mathrm{TC}(S) = S \cup \bigcup S \cup \bigcup \bigcup S \cup \dots$.]

2. Prove the following stronger version of Regularity: If C is a nonempty class, then for some $x \in C$, $x \cap C = 0$.

[*Hint*: Let $S \in C$, $S \cap C \neq 0$. Apply A8 to the set $\mathrm{TC}(S) \cap C$.]

3. Use Problem 2 to prove the $\in$-induction.

4. Prove the following version of $\in$-recursion: If C is a class, then there is a unique class HC such that

$$x \in \mathrm{H}C \quad \text{iff} \quad x \in C \text{ and } x \subseteq \mathrm{H}C.$$

5. Prove that $V = \bigcup_{\alpha \in \mathrm{On}} V_\alpha$, using the $\in$-induction.

6. Use Regularity to show that the following principle implies Replacement:

$$\forall X \, \exists Y \, (\forall u \in X) \, [\exists v \, \phi(u, v, \vec{p}) \rightarrow (\exists v \in Y) \, \phi(u, v, \vec{p})]$$

7. *Reflection Principle*. Let ϕ be a formula. For each set S_0, there is a set $S \supseteq S_0$ such that for all $\vec{x} \in S$,

$$\phi(\vec{x}) \quad \text{iff} \quad S \models \phi(\vec{x}),$$

[*Hint*: Apply Lemma 3.3 to all subformulas of ϕ.]

8. Using the Axiom of Choice, show that the set S in Lemma 3.3 can be found such that $|S| = |S_0|$ (provided S_0 is infinite).

[*Hint*: In the proof, choose one element in each $\hat{C}_y$.]

9. Using the Axiom of Choice, prove the stronger version of the Reflection Principle which requires that $|S| = |S_0|$ (provided S_0 is infinite).

[*Hint*: Use Problem 8.]

*10. There is a theorem Θ of ZF such that the formula 'x is constructible' is absolute for every transitive $\mathscr{M}$ which satisfies Θ.

[*Hint*: Θ says that $\mathscr{M}$ has the closure properties mentioned in (a)–(j) in the proof of Theorem (3.5).]

*11. $2^{\aleph_0} = \aleph_1$ holds in L.

[*Hint*: Assume $V = L$, and prove that if $X \subseteq \omega$, then $X \in L_\alpha$ for some $\alpha < \omega_1$. Let β be such that $X \in L_\beta$. By the Reflection Principle, there is a countable S containing all $n \in \omega$, X and β, satisfying Extensionality and Θ. Let M be the transitive isomorph of S. $X \in L_\alpha$ for some $\alpha \in M$; α is countable.]

*12. L satisfies the Generalized Continuum Hypothesis ($2^{\aleph_\alpha} = \aleph_{\alpha+1}$ for every α).

[*Hint*: Generalize Problem 11.]

Let A be a set of ordinals. Let

$$L[A] = \bigcup_{\alpha \in On} L_\alpha[A],$$

where

$$L_0[A] = 0,$$
$$L_\alpha[A] = \bigcup_{\beta<\alpha} L_\beta[A] \quad \text{if } \alpha \text{ is limit,}$$
$$L_{\alpha+1}[A] = \mathscr{P}(L_\alpha[A]) \cap \text{cl}(L_\alpha[A] \cup \{L_\alpha[A]\} \cup \{L_\alpha[A] \cap A\}).$$

13. Show that:
(a) $A \in L[A]$;
(b) $L[A] \models (V = L[A])$;
(c) $L[A]$ is a model of ZF+Axiom of Choice;
(d) $L[A]$ is the least model of ZF which contains all ordinals and A.

14. Show that $V_\omega = L_\omega$ is a model of ZF−Infinity.

15. Let $\mathscr{M}$ be a transitive model of ZF. Show that:
(a) $\mathscr{P}^{\mathscr{M}}(X) = \mathscr{P}(X) \cap \mathscr{M}$;
(b) $V_\alpha^{\mathscr{M}} = V_\alpha \cap \mathscr{M}$;
(c) if $\mathscr{M} \models |X| = |Y|$, then $|X| = |Y|$;
(d) if α is a cardinal (resp. regular cardinal, singular cardinal), then $\mathscr{M} \models \alpha$ is a cardinal (resp. regular cardinal, singular cardinal).

A relation ρ on a set A is *well-founded* if for every nonempty $X \subseteq A$ there is $u \in X$ such that there is no $v \in X$ with $x\rho u$.

16. A relation ρ on A is well-founded if and only if there is a mapping π from A to ordinals such that $x\rho y$ implies $\pi x < \pi y$.

[*Hint*: Use transfinite or $\in$-recursion.]

17. Using the Principle of Dependent Choices, show that ρ is well-founded if and only if there is no infinite sequence $x_0, x_1, x_2, \ldots$ such that $x_1 \rho x_0$, $x_2 \rho x_1$, etc..

18. The formula 'ρ is well-founded' is absolute for models of ZF.

[*Hint*: Use Problem 16. The formula can be written either as $\forall X\, \phi(X, \rho)$ or as $\exists \pi\, \psi(\pi, \rho)$, where ϕ and ψ are absolute.]

*19. Let $\mathcal{M}, \mathcal{N}$ be transitive models of ZF which have the same sets of ordinals. If one of them satisfies the Axiom of Choice, then $\mathcal{M} = \mathcal{N}$.

[*Hint*: $\mathcal{M}$ and $\mathcal{N}$ have the same subsets of $\mathrm{On} \times \mathrm{On}$. Assume $\mathcal{M} \models$ Axiom of Choice. First show that $\mathcal{M} \subseteq \mathcal{N}$. If $X \in \mathcal{M}$, then $\in$ on $\mathrm{TC}(\{X\})$ can be represented by a well-founded relation ρ on ordinals. By assumption, $\rho \in \mathcal{N}$, and since well-founded, it is isomorphic to a transitive set which must be $\mathrm{TC}(\{X\})$ and so $X \in \mathcal{N}$. To show $\mathcal{N} \subseteq \mathcal{M}$, use the $\in$-induction. Prove that $X \in \mathcal{N}$, $X \subseteq Y$ and $Y \in \mathcal{M}$ implies $X \in \mathcal{M}$; this uses the Axiom of Choice in $\mathcal{M}$.]

For any set X and any natural number n, define $\mathcal{P}^n(X)$ as follows:

$$\mathcal{P}^0(X) = X, \qquad \mathcal{P}^{n+1}(X) = \mathcal{P}(\mathcal{P}^n(X)).$$

Consider the following statements:

K(n) *For every set S there exists an ordinal α and a one-to-one function f which maps S into $\mathcal{P}^n(\alpha)$.*

K(0) is equivalent to the Axiom of Choice. K(1) is equivalent to the *Selection Principle* (Problem 4.12).

The following result is a generalization of Problem 19.

*20. Let n be a natural number. Let $\mathcal{M}, \mathcal{N}$ be transitive models of ZF such that $(\mathcal{P}^{n+1}(\alpha))^{\mathcal{M}} = (\mathcal{P}^{n+1}(\alpha))^{\mathcal{N}}$ for every ordinal α. If one of the models satisfies K(n), then $\mathcal{M} = \mathcal{N}$.

The class of all *ordinally definable* sets, OD, is the closure under Gödel operations of the class of all V_α, $\alpha \in \mathrm{On}$.

*21. The ordinally definable sets are exactly all the sets of the form

$$X = \{u : \phi(u, \alpha_1, \ldots, \alpha_n)\},$$

where $\alpha_1, \ldots, \alpha_n$ are ordinals.

[*Hint*: As in Theorem 3.2, one can prove that X can be obtained by application of Gödel operations to $V_{\alpha_1}, \ldots, V_{\alpha_n}$ and additional parameters $V_{\gamma_1}, \ldots, V_{\gamma_k}$.]

22. There is a class HOD such that $x \in \mathrm{HOD}$ if and only if $x \in \mathrm{OD}$ and $x \subseteq \mathrm{HOD}$.

[*Hint*: Use Problem 4.]

*23. HOD is a model of ZF + Axiom of Choice.

[*Hint*: To show that HOD is almost universal, prove that $V_\alpha \cap \text{HOD} \in \text{OD}$; use Problem 21.]

This gives another consistency proof of the Axiom of Choice. Caution: 'x is ordinally definable' is not absolute.

3.6. Historical remarks

The consistency of the Axiom of Choice and of the Generalized Continuum Hypothesis was proved by Gödel [1938, 1939, 1940], who also invented constructible sets. Other contributors to the theory of transitive models include Shepherdson [1951] (transitive models), Mostowski [1949] (the Isomorphism Theorem), Montague [1961] (the Reflection Principle) and Levy [1957] ($L[A]$). The result in Problem 19 was proved by Vopěnka and Balcar [1967]; the generalisation in Problem 20 is due to Monro [1972]. The basic results about ordinally definable sets are due to Vopěnka and Balcar [1967] and Myhill and Scott [1967].

CHAPTER 4

PERMUTATION MODELS

4.1. Set theory with atoms

Having established the consistency of the Axiom of Choice, our next goal will be to show that the Axiom of Choice is *independent* from the other axioms; that is, to show that it is consistent to assume that the Axiom of Choice fails.

The independence of the Axiom of Choice will be the main result of Chapter 5. In the present chapter, we describe an older method which, although it does not solve the problem of independence of the Axiom of Choice in the ordinary set theory, sheds some light on the problem by establishing its independence in the axiomatic set theory with atoms.

The *set theory with atoms*, ZFA, is a modified version of set theory, and is characterized by the fact that it admits objects other than sets, *atoms.* Atoms are objects which do not have any elements.

The language of ZFA consists of $=$ and $\in$ and of two constant symbols 0 and A (the empty set and the set of all atoms). The axioms of ZFA are like the axioms of ZF, except for the following changes:

0. *Empty set* $\qquad \neg \exists x\,(x \in 0)$.

A. *Atoms* $\qquad \forall z\,[z \in A \leftrightarrow z \neq 0 \wedge \neg \exists x\,(x \in z)]$.

Atoms are the elements of A; *sets* are all objects which are not atoms.

A1. *Extensionality*

$$(\forall \text{ set } X)\,(\forall \text{ set } Y)\,[\forall u\,(u \in X \leftrightarrow u \in Y) \leftrightarrow X = Y].$$

A8. *Regularity*

$$(\forall \text{ nonempty } S)\,(\exists x \in S)\,[x \cap S = 0].$$

Note that 'X is nonempty' is not the same as '$X \neq 0$'; it is the same only if X is a set. Some operations make sense only for sets, e.g., $\bigcup X$ or $\mathscr{P}(X)$, some also for atoms, like $\{x, y\}$. If we add to ZFA the axiom $A = 0$, then we get ZF. However, we are now more interested in the case $A \neq 0$. Incidentally, the theory

$$\text{ZFA} + A \text{ is infinite}$$

is consistent, provided ZF is consistent (see Problem 1).

The development of the theory ZFA is very much the same as that of ZF. (Caution: In the definition of ordinals, we have to insert the clause that an ordinal does not have atoms among its elements.) One can define the *rank* of sets and build a hierarchy analogous to the V_α's. For any set S, let $\mathscr{P}^\alpha(S)$ be defined as follows:

$$\mathscr{P}^0(S) = S,$$
$$\mathscr{P}^{\alpha+1}(S) = \mathscr{P}^\alpha(S) \cup \mathscr{P}(\mathscr{P}^\alpha(S)),$$
$$\mathscr{P}^\alpha(S) = \bigcup_{\beta<\alpha} \mathscr{P}^\beta(S) \qquad (\alpha \text{ limit})$$

and let

$$\mathscr{P}^\infty(S) = \bigcup_{\alpha \in \text{On}} \mathscr{P}^\alpha(S).$$

Then we have

$$V = \mathscr{P}^\infty(A).$$

The class $\mathscr{P}^\infty(0)$ is a model of ZF and is called the *kernel*. Note that all the ordinals are in the kernel.

A transitive set does not necessarily contain 0 and may have nontrivial automorphisms (e.g., the set $\{a_1, a_2\}$, where $a_1, a_2 \in A$). A transitive class which is almost universal and closed under Gödel operations is a model of ZFA, provided it contains 0 (that is, when we want to interpret $\in$ as $\in$, 0 as 0, and atoms as atoms).

4.2. Permutation models

The underlying idea of permutation models is the fact that the axioms of ZFA do not distinguish between the atoms, and that they are used to construct models in which the set A has no well-ordering.

Let π be a permutation of the set A. Using the hierarchy of $\mathscr{P}^\alpha(A)$'s, we can define πx for every x as follows:

$$\pi(0) = 0, \qquad \pi(x) = \pi''x = \{\pi(y) : y \in X\}$$

(either by $\in$-recursion or by recursion on the rank of x). Under this definition, π becomes an $\in$-automorphism of the universe and one can easily verify the following facts about π:

(a). $x \in y \leftrightarrow \pi x \in \pi y$.
(b). $\phi(x_1, ..., x_n) \leftrightarrow \phi(\pi x_1, ..., \pi x_n)$.
(c). rank $(x) =$ rank (πx).
(d). $\pi\{x, y\} = \{\pi x, \pi y\}$, $\pi(x, y) = (\pi x, \pi y)$.
(e). If R is a relation, then πR is a relation and $(x, y) \in R \leftrightarrow (\pi x, \pi y) \in R$.
(f). If f is a function on X, then πf is a function on πX and $(\pi f)(\pi x) = \pi(f(x))$.
(g). $\pi x = x$ for every x in the kernel.
(h). $(\pi \cdot \rho)x = \pi(\rho(x))$.

Let $\mathscr{G}$ be a group of permutations of A. A set $\mathscr{F}$ of subgroups of $\mathscr{G}$ is a *normal filter* on $\mathscr{G}$ if for all subgroups H, K of $\mathscr{G}$:

(i) $\mathscr{G} \in \mathscr{F}$;
(ii) if $H \in \mathscr{F}$ and $H \subseteq K$, then $K \in \mathscr{F}$;
(iii) if $H \in \mathscr{F}$ and $K \in \mathscr{F}$, then $H \cap K \in \mathscr{F}$;
(iv) if $\pi \in \mathscr{G}$ and $H \in \mathscr{F}$, then $\pi H \pi^{-1} \in \mathscr{F}$;
(v) for each $a \in A$, $\{\pi \in \mathscr{G} : \pi a = a\} \in \mathscr{F}$.

For each x, let

$$\text{sym}_{\mathscr{G}}(x) = \{\pi \in \mathscr{G} : \pi x = x\};$$

$\text{sym}_{\mathscr{G}}(x)$ is a subgroup of $\mathscr{G}$.

Let $\mathscr{G}$ and $\mathscr{F}$ be fixed. We say that x is *symmetric* if sym $(x) \in \mathscr{F}$. The class

$$\mathscr{V} = \{x : x \text{ is symmetric and } x \subseteq \mathscr{V}\}$$

consists of all *hereditarily symmetric* objects (cf. Problem 3.4); we call $\mathscr{V}$ a *permutation model.*

THEOREM 4.1. *$\mathscr{V}$ is a transitive model of* ZFA; *$\mathscr{V}$ contains all the elements of the kernel and also $A \in \mathscr{V}$.*

PROOF. Obviously, $\mathscr{V}$ is transitive. To see that $\mathscr{V}$ is closed under Gödel operations, it suffices to show that for all x, y,

$$\text{sym}\,(\mathscr{F}_i(x, y)) \supseteq \text{sym}\,(x) \cap \text{sym}\,(y), \qquad i = 1, ..., 8.$$

We leave it to the reader to verify this. To show that $\mathscr{V}$ is almost universal, let us prove that for each α, the set $\mathscr{P}^{\alpha}(A) \cap \mathscr{V}$ is symmetric; actually, we show that sym $(\mathscr{P}^{\alpha}(A) \cap \mathscr{V}) = \mathscr{G}$. It is easy to see that for all x,

$$\text{sym}\,(\pi x) = \pi\,\text{sym}\,(x) \cdot \pi^{-1};$$

thus if x is symmetric and $\pi \in \mathscr{G}$, then πx is symmetric, and so, by induction, if $x \in \mathscr{V}$, then $\pi x \in \mathscr{V}$ for all x, and all $\pi \in \mathscr{G}$. Since rank (πx) = rank (x), it follows that $\pi(\mathscr{P}^\alpha(A) \cap \mathscr{V}) = \mathscr{P}^\alpha(A) \cap \mathscr{V}$ for all α, all $\pi \in \mathscr{G}$. Thus $\mathscr{V}$ is transitive, almost universal and closed under Gödel operations. It remains to be shown that $\mathscr{P}^\infty(0) \subseteq \mathscr{V}$ and $A \in \mathscr{V}$. The former is true because sym $(x) = \mathscr{G}$ for all $x \in \mathscr{P}^\infty(0)$, and the latter is true because sym $(A) = \mathscr{G}$ and sym $(a) \in \mathscr{F}$ for each $a \in A$.

Most of the permutation models we will consider will be of the following simple type: Let $\mathscr{G}$ be a group of permutations of A. A family I of subsets of A is a *normal ideal* if for all subsets E, F of A:

(4.1) (i) $0 \in I$;
(ii) if $E \in I$ and $F \subseteq X$, then $F \in I$;
(iii) if $E \in I$ and $F \in I$, then $E \cup F \in I$;
(iv) if $\pi \in \mathscr{G}$ and $E \in I$, then $\pi''E \in I$;
(v) for each $a \in A$, $\{a\} \in I$.

For each x, let

$$\text{fix}_{\mathscr{G}}(x) = \{\pi \in \mathscr{G} : \pi y = y \text{ for all } y \in x\};$$

$\text{fix}_{\mathscr{G}}(x)$ is a subgroup of $\mathscr{G}$.

Let $\mathscr{F}$ be the filter on $\mathscr{G}$ generated by the subgroups fix (E), $E \in I$. $\mathscr{F}$ is a normal filter, and so it defines a permutation model $\mathscr{V}$. Note that x is symmetric if and only if there exists $E \in I$ such that

$$\text{fix}(E) \subseteq \text{sym}(x).$$

We say that E is a *support* of x.

Now, everything is ready to start with applications of permutation models. Before doing so, one more remark will be useful. When constructing permutation models, we shall always work in the theory ZFA+Axiom of Choice. (For the consistency, see Problem 1). If $\mathscr{V}$ is a permutation model, $\mathscr{V}$ contains all elements of the kernel and so the Axiom of Choice holds in the kernel. In particular, every $x \in \mathscr{P}^\infty(0)$ can be well-ordered. Therefore any $x \in \mathscr{V}$ can be well-ordered if and only if there is a one-to-one mapping f of x into the kernel. For any such f, however, $\pi f = f$ if and only if $\pi \in$ fix (x). Thus we have

(4.2) $\mathscr{V} \models$ (x can be well-ordered) iff fix $(x) \in \mathscr{F}$.

4.3. The basic Fraenkel model

We present a simple example of a permutation model that does not satisfy the Axiom of Choice.

Assume that the set A is countable.[1] Let $\mathscr{G}$ be the group of all permutations of A, and let I be the set of all finite subsets of A. Obviously, I satisfies the conditions (1.1); let $\mathscr{V}$ be the corresponding permutation model. Note that x is symmetric if and only if there is a finite $E \subseteq A$ such that $\pi x = x$ whenever $\pi a = a$ for all $a \in E$.

It is easy to see that the subgroup fix (A) is not in the filter generated by $\{\text{fix}(E) : E \subseteq A \text{ finite}\}$: For every finite $E \subset A$, one can easily find $\pi \in \mathscr{G}$ such that $\pi \in \text{fix}(E)$ and $\pi \notin \text{fix}(A)$. By (4.2), it follows that the set has no well-ordering in the model $\mathscr{V}$. Thus we get:

THEOREM 4.2. *The Axiom of Choice is unprovable in set theory with atoms.*

4.4. The second Fraenkel model

In this section, we shall construct a model in which the Axiom of Choice fails even for countable families of pairs.

Assume that the set A is countable and divide it into countably many disjoint pairs:

$$A = \bigcup_{n=0}^{\infty} P_n, \qquad P_n = \{a_n, b_n\}, \quad n = 0, 1, \ldots.$$

Let $\mathscr{G}$ be the group of all those permutations of A which preserve the pairs P_n, i.e.,

$$\pi(\{a_n, b_n\}) = \{a_n, b_n\}, \qquad n = 1, 2, \ldots.$$

Let I be the ideal of finite subsets of A. Clearly, I is a normal ideal. A set x is symmetric iff there is k such that $\pi x = x$ whenever $\pi \in \mathscr{G}$ and

$$\pi a_0 = a_0, \quad \pi b_0 = b_0, \quad \ldots, \quad \pi a_k = a_k, \quad \pi b_k = b_k.$$

Let $\mathscr{V}$ be the permutation model determined by $\mathscr{G}$ and I. Then $\mathscr{V}$ has the following properties:

(a). Each P_n is in $\mathscr{V}$.

(b) The sequence $\langle P_n : n \in \omega \rangle$ is in $\mathscr{V}$; thus the set $\{P_n : n \in \omega\}$ is countable in $\mathscr{V}$.

(c) There is no function $f \in \mathscr{V}$ such that dom $f = \omega$ and $f(n) \in P_n$ for each n.

[1] *Countable* means always 'infinite countable'.

Proof. (a). $\pi P_n = P_n$ for each n and each $\pi \in \mathscr{G}$.

(b). $\pi(\langle P_n : n \in \omega \rangle) = \langle P_n : n \in \omega \rangle$ for each $\pi \in \mathscr{G}$.

(c). Assume that f is such a function and $\{a_0, b_0, \ldots, a_k, b_k\}$ is its support. Let $\pi \in \text{fix}\,\{a_0, b_0, \ldots, a_k, b_k\}$ and $\pi a_{k+1} = b_{k+1}$. Then $\pi f = f$, $\pi(k+1) = k+1$ and $\pi(f(k+1)) \neq f(k+1)$, a contradiction.

Thus in the model $\mathscr{V}$ there is no choice function on the countable family $\{P_n : n \in \omega\}$ and we have the following:

THEOREM 4.3. *The Axiom of Choice for countable families of pairs is unprovable in set theory with atoms.*

4.5. The ordered Mostowski model

In both examples in Sections 4.3 and 4.4, the Axiom of Choice is violated in a very strong way; in each case there exists a family of finite sets which does not have a choice function (see Problem 3). The present example shows how to violate the general Axiom of Choice and still preserve a weaker version; in this case, the Ordering Principle (and a fortiori the Axiom of Choice for families of finite sets).

First, a few words about permutation models in general: Let $\mathscr{V}$ be a permutation model (given by $\mathscr{G}$ and $\mathscr{F}$) and let $C \subseteq \mathscr{V}$ be a class. We say that C is *symmetric* if $\text{sym}\,(C) \in \mathscr{F}$, where

$$\text{sym}\,(C) = \{\pi \in \mathscr{G} : \pi''C = C\}.$$

If for each α, $C_\alpha = C \cap \mathscr{P}^\alpha(A)$ is the set of all $x \in C$ of rank less than α, we notice that if C is symmetric, then $C_\alpha \in \mathscr{V}$ for each α (and $C \cap x \in \mathscr{V}$ for each $x \in \mathscr{V}$).

If one works in a set theory which admits classes, like the Gödel–Bernays set theory, then one can consider symmetric classes as classes of the permutation model. In our case, when we use classes informally, symmetric classes serve as a canonical way of construction of sets in the model $\mathscr{V}$.

LEMMA 4.4. *Let $\mathscr{G}$ be a group of permutations of A and let I be a normal ideal of supports; let $\mathscr{V}$ be the permutation model given by $\mathscr{G}$ and I. Then the class of all pairs (E, x) such that $E \in I$, $x \in \mathscr{V}$ and E is a support of x, is symmetric.*

PROOF. If $\pi \in \mathscr{G}$, then

$$\text{fix}\,(\pi E) = \pi\,\text{fix}\,(E) \cdot \pi^{-1},$$
$$\text{sym}\,(\pi x) = \pi \cdot \text{sym}\,(x) \cdot \pi^{-1}.$$

Thus if E is a support of x, then πE is a support of πx.

Now we shall construct the Mostowski model. Assume that the set A is countable, and let $<$ be an ordering of A such that A is densely ordered and without smallest or greatest element (thus A is isomorphic to the rationals). Let $\mathscr{G}$ be the group of all order-preserving permutations of A, and let I be the ideal of finite subsets of A.

Let $\mathscr{V}$ be the permutation model given by $\mathscr{G}$ and I. We shall show that $\mathscr{V}$ has the following properties:

(a). The set A cannot be well-ordered in $\mathscr{V}$.

(b). There is a linear ordering $\prec$ of $\mathscr{V}$ which is a symmetric class.

It is easy to see that $\operatorname{fix} E \nsubseteq \operatorname{fix} A$ for any finite $E \subset A$, and so we have (a).

To prove (b), we need several lemmas. The first one is crucial and shows that this model $\mathscr{V}$ has the important feature that every $x \in \mathscr{V}$ has a *least* support.

LEMMA 4.5

(a). *If E_1 is a support of x and E_2 is a support, then $E = E_1 \cap E_2$ is a support of x.*

(b). *Every symmetric x has a least support. The class of all pairs (x, E), where $x \in \mathscr{V}$ and E is the least support of x, is symmetric.*

PROOF. (a). The proof is based on the following fact, which can be proved more easily by drawing a picture than in writing, and thus we leave it to reader's imagination: Let E_1 and E_2 be finite subsets of A, $E = E_1 \cap E_2$. If $\pi \in \mathscr{G}$ is such that $\pi(a) = a$ for each $a \in E$, then we can find a finite number of permutations $\pi_1 \rho_1, \ldots, \pi_n \rho_n$ such that $\pi_i \in \operatorname{fix}(E_1)$ and $\rho_i \in \operatorname{fix}(E_2)$ for each $i = 1, \ldots, n$, and $\pi = \pi_i \rho_i \ldots \pi_n \rho_n$. Thus $\operatorname{fix}(E) = [\operatorname{fix}(E_n) \cup \operatorname{fix}(E_2)]$, where $[X]$ denotes the group generated by X. In particular, if both E_1 and E_2 are supports of x and $\pi \in \operatorname{fix}(E)$, then $\pi x = \pi_i \rho_i \ldots \pi_n \rho_n x = x$ and so E is a support of x.

(b). The least support of x is the intersection of all supports of x. Since their number is finite, the intersection is a support of x by (a). The rest is as in Lemma 4.4: If E is the least support of x, then πE is the least support of πx.

LEMMA 4.6

(a). *If E is a support of x and $\pi E = E$, then $\pi x = x$.*

(b). *There is a symmetric class F which is a one-to-one function of $\mathscr{V}$ into* $\mathrm{On} \times I$.

PROOF. (a). Since every $\pi \in \mathscr{G}$ is order-preserving, it follows that if $\pi E = E$, then $\pi \in \text{fix}(E)$.

(b). For each $x \in \mathscr{V}$, let orb (x) be the orbit of x:

$$\text{orb}(x) = \{\pi x : \pi \in \mathscr{G}\}.$$

The orbits are pairwise disjoint and sym (orb (x)) $= \mathscr{G}$ for each x. Thus if we enumerate the orbits by ordinals, the enumeration is a symmetric class. Let $F_1(x)$ be the number of orb (x) and let $F_2(x)$ be the least support of x, for each $x \in \mathscr{V}$. Let $F(x) = (F_1(x), F_2(x))$; obviously, F is symmetric. To see that it is one-to-one, note that if x and πx are in the same orbit and if E is the least support of x, then πE is the least support of πx, and we have $x = \pi x$ if and only if $E = \pi E$ by (a).

The rest is easy. The ordering $<$ of A is in $\mathscr{V}$ since the group $\mathscr{G}$ consists of order-preserving permutations and so sym $(<) = \mathscr{G}$. The set I consists of finite subsets of a linearly ordered set and thus can be linearly ordered (lexicographically). Therefore the class $\text{On} \times I$ can be linearly ordered (lexicographically again). And since we have a symmetric one-to-one mapping of $\mathscr{V}$ into $\text{On} \times I$, we obtain a linear ordering $\prec$ of $\mathscr{V}$, which is a symmetric class. Thus every set can be linearly ordered in $\mathscr{V}$. (As a consequence, every family of finite sets has a choice function.) This constitutes the proof of the following theorem.

THEOREM 4.7. *The Axiom of Choice is independent from the Ordering Principle in set theory with atoms.*

4.6. Problems

1. The theory (ZFA + Axiom of Choice + A is infinite) is consistent relative to ZF + Axiom of Choice.

[*Hint*: Let C be an infinite set of infinite subsets of ω. Construct a model $\Pi = \bigcup_{\alpha \in \text{On}} \Pi_\alpha$, where

$$\Pi_0 = C, \qquad \Pi_{\alpha+1} = \Pi_\alpha \cup \mathscr{P}(\Pi_\alpha) - \{0\}.$$

Choose $a_0 \in C$, and interpret it as the empty set, and interpret $C - \{a_0\}$ as atoms. If $X \in \Pi$, then each well-ordering $< \subseteq X \times X$ is also in Π.]

2. If x can be well-ordered in $\mathscr{V}$, then $\mathscr{P}(x)$ can be well-ordered in $\mathscr{V}$.
[*Hint*: Use (4.2).]

3. In the basic Fraenkel model, the family $S = \{\{a, b\} : a, b \in A\}$ has no choice function. Consequently, A cannot be linearly ordered.

4. In each of the three models constructed in this chapter, the set A is Dedekind finite (D-finite).
[*Hint*: X is D-infinite iff $|X| \geqslant \aleph_0$.]

5. In both the basic Fraenkel model and the Mostowski model, the set $\mathscr{P}(A)$ is Dedekind finite.

[*Hint*: For every finite $E \subset A$, there are only finitely many subsets of A which have E as a support.]

6. For every infinite set X, $\mathscr{P}(\mathscr{P}(X))$ is D-infinite.
[*Hint*: Consider the set $\{S_n\}_{n=0}^{\infty}$, where $S_n = \{Y \subset X : |Y| = n\}$.]

Call a set *amorphous* iff it is infinite but is not a union of two disjoint infinite sets.

7. In the basic Fraenkel model, the set A is amorphous.

8. A set S is finite if and only if every nonempty $X \subseteq \mathscr{P}(S)$ has a $\subseteq$-maximal element.
[*Hint*: Consider the set $X = \{x \subseteq S : x \text{ is finite}\}$.]

Call a set S T-*finite* if every nonempty monotone $X \subseteq \mathscr{P}(S)$ has a $\subseteq$-maximal element.

9. Every finite set is T-finite. Every T-finite set is D-finite.
[*Hint*: S is D-finite iff $|S| \not\geqslant \aleph_0$.]

10. Every amorphous set is T-finite.
[*Hint*: Let S be amorphous and T-infinite; let X be a $\subseteq$-chain in $\mathscr{P}(T)$ without a maximal element. Show that the order-type of X must be ω and derive a contradiction by dividing ω into even and odd numbers.]
This together with Problem 7 shows that one cannot prove that every T-finite set is finite.

11. Every infinite linearly ordered set is T-infinite.
[*Hint*: Consider the $\subseteq$-chain $\{S_x : x \in S\}$, where $S_x = \{y \in S : y \leqslant x\}$.]
This together with Problem 4 shows that we cannot prove that every D-finite set is T-finite.

Consider the following statement:

SELECTION PRINCIPLE. *For every family $\mathscr{F}$ of sets, each containing at least two elements, there is a function f such that*

$$0 \neq f(S) \subsetneqq S$$

for each set S in the family $\mathscr{F}$. (We say that f is a *selector* on $\mathscr{F}$.)

12. The Selection Principle is equivalent to the following statement (K(2) of Problem 3.20):

For every set M there exists an ordinal α and a one-to-one function g which maps M into $\mathscr{P}(\alpha)$.

Consequently, the Selection Principle implies the Ordering Principle.

[*Hint*: (a). Assume that the Selection Principle is true. Let M be a set and let f be a selector on $\mathscr{P}(M)$. By recursion on the length of the sequences, construct sets M_t, where t ranges over transfinite sequences of 0's and 1's: $M_0 = f(M)$, $M_1 = M - M_0$, $M_{t^\frown 0} = f(M_t)$, $M_{t^\frown 1} = M_t - M_{t^\frown 0}$, and $M_t = \bigcap_{s \subset t} M_s$ for t of limit length. For some α, all M_t's are either empty or singletons, for all $t \in {}^{\alpha}2$. This gives a one-to-one mapping of M into ${}^{\alpha}2$.

(b). Assume that the statement is true. It suffices to prove the Selection Principle for families of sets which are subsets of some $\mathscr{P}(\alpha)$. Let $\mathscr{F}$ be such a family. For each $S \in \mathscr{F}$, $S \subseteq \mathscr{P}(\alpha)$, let β be least such that $\beta \in X$ for some $X \in S$ and $\beta \notin Y$ for some $Y \in S$. Let $f(S) = \{X \in S : \beta \in X\}$; then f is a selector on $\mathscr{F}$.]

13. The Selection Principle fails in every permutation model in which the set A cannot be well-ordered.

[*Hint*: A one-to-one mapping into the kernel yields a well-ordering.]

Thus the Selection Principle is independent from the Ordering Principle in set theory with atoms.

14. The following version of the Axiom of Choice holds in the Mostowski model:

For every family of nonempty well-orderable sets there exists a choice function.

[*Hint*: For each $E \in I$, let $\Delta(E)$ be the class of all $x \in \mathscr{V}$ such that x is a support of x. The function $F : \mathscr{V} \to \mathrm{On} \times I$ from Lemma 4.7 restricted to $\Delta(E)$ has values in $\mathrm{On} \times \mathscr{P}(E)$. Since $\mathscr{P}(E)$ is finite, the ordering $\prec$ of $\mathscr{V}$ is a well-ordering on $\Delta(E)$. If $S \in \mathscr{V}$ is well-orderable, then for some $E \in I$, fix $(E) \subseteq$ fix (S) and so $S \subseteq \Delta(E)$; hence $\prec$ well-orders S. Choose the $\prec$-least element in each well-orderable S.]

4.7. Historical remarks

The method of permutation models was introduced by Fraenkel [1922–37] and, in a precise version (with supports) by Mostowski [1938–39]. The present version (with filters) is due to Specker [1957]. Theorems 4.2 and 4.3 are Fraenkel's; for Theorem 4.7, see Mostowski [1939].

The results of Problems 4–11 are in Lindenbaum and Mostowski [1938], and Mostowski [1938] and Levy [1958]. The equivalence in Problem 8 and the definition of T-finiteness are due to Tarski [1924b]. The Selection Principle was formulated by Kuratowski [1922] and the equivalence in Problem 12 was proved by Kinna and Wagner [1955]; the observation in Problem 13 is due to Mostowski [1958]. The result of Problem 14 is by Läuchli [1964]. For additional contributions to the method of permutation models, cf. Doss [1945], Mendelson [1948, 1956], Jesenin-Vol'pin [1954], Shoenfield [1955] and Fraïssé [1958].

CHAPTER 5

INDEPENDENCE OF THE AXIOM OF CHOICE

5.1. Generic models

By the result of Chapter 3, it is consistent to assume that $V = L$, i.e., that the universe itself is the only transitive proper class which is a model of set theory. Thus if we want to construct models of ZF in which the Axiom of Choice fails, we have to look for extensions of the universe, rather than its submodels.

In the present section, we describe a general method of extending a given transitive model $\mathcal{M}$ to a model $\mathcal{M}[G]$, a *generic* model; the following section will deal with certain submodels of generic models, and these will be used in subsequent sections to prove the independence of the Axiom of Choice.

We start with introducing so-called *Boolean-valued models*. These are a generalization of ordinary models in the sense that the truth-values are not 0 and 1 (false or true), but are elements of a given complete Boolean algebra. Since the Boolean-valued models will be subsequently used to obtain generic extensions of transitive models, we will use the following more general framework: Instead of working in the universe, we fix a transitive model $\mathcal{M}$ (the *ground model*) and will carry out the construction of a Boolean-valued model *inside* $\mathcal{M}$. Moreover, we shall always assume that $\mathcal{M}$ *satisfies the Axiom of Choice*.

Let B be a Boolean algebra with operators $+$, $\cdot$ and $-$, and the corresponding partial ordering $\leqslant$. The algebra B is *complete* if every nonempty subset of B has a least upper bound and a greatest lower bound. Then we introduce infinitary operators $\sum$ and $\prod$ (sum and product) as follows:

$$\sum \{a : a \in A\} = \text{lub}(A), \qquad \prod\{a : a \in A\} = \text{glb}(A).$$

For our convenience, let us introduce another operation on a Boolean algebra:

$$u \Rightarrow v = -u+v$$

(compare with $\phi \to \psi \leftrightarrow \neg \phi \vee \psi$).

5.1.1. *Boolean-valued models*

Let $\mathcal{M}$ be a transitive model of ZF+AC. Until further notice, we will be working inside $\mathcal{M}$. Let $B \in \mathcal{M}$ be a fixed complete Boolean algebra (that is, complete in $\mathcal{M}$).[1] We define a class $\mathcal{M}^B$, a Boolean-valued model, by recursion:

$$\mathcal{M}^B = 0,$$

$$\mathcal{M}^B_\alpha = \bigcup_{\beta<\alpha} \mathcal{M}^B_\beta \text{ if } \alpha \text{ is a limit ordinal;}$$

$$\mathcal{M}^B_{\alpha+1} = \{x: x \text{ is a function, dom } (x) \subseteq \mathcal{M}^B_\alpha \text{ and rng } (x) \subseteq B\};$$

$$\mathcal{M}^B = \bigcup_{\alpha \in \mathrm{On}} \mathcal{M}^B_\alpha.$$

There is a natural embedding $\check{}$ of $\mathcal{M}$ in $\mathcal{M}^B$. We define $\check{}$ by the $\in$-recursion:

$\check{0} = 0,$

$\check{x}$ is the element of $\mathcal{M}^B$ whose domain is $\{\check{y} : y \in x\}$ and $\check{x}(\check{y}) = 1$ for all $y \in x$.

For every formula $\phi(x_1, ..., x_n)$ with variables in $\mathcal{M}^B$, we shall define its Boolean value $[\![\phi]\!]$ which will be an element of B. First, we give (a somewhat involved) definition of the *Boolean values* $[\![x \in y]\!]$ and $[\![x = y]\!]$ for all $x, y \in \mathcal{M}^B$. The definition is by recursion on pairs of ordinals $(\rho(x), \rho(y))$, in the canonical well-ordering of On × On, where $\rho(x)$ denotes the least α such that $x \in \mathcal{M}^B_{\alpha+1}$. We let

$$[\![x \in y]\!] = \sum_{z \in \mathrm{dom}(y)} (y(z) \cdot [\![z = x]\!]),$$

$$[\![x = y]\!] = \prod_{z \in \mathrm{dom}(x)} (x(z) \Rightarrow [\![z \in y]\!]) \cdot \prod_{z \in \mathrm{dom}(y)} (y(z) \Rightarrow [\![z \in x]\!]).$$

(Compare with

$$x \in y \leftrightarrow (\exists z \in y)[z = x]$$

and

$$x = y \leftrightarrow (\forall z \in x)[z \in y] \wedge (\forall z \in y)[z \in x].)$$

[1] Being a Boolean algebra is an absolute property. On the other hand, $B \in \mathcal{M}$ may be complete in $\mathcal{M}$ but not in the universe, since the former means only that l.u.b. exists for every subset of B which is in $\mathcal{M}$.

Once we have defined the Boolean values $[\![x \in y]\!]$ and $[\![x = y]\!]$, we can define the Boolean values for any formula, by recursion on its complexity:

$$[\![\neg \phi]\!] = -[\![\phi]\!],$$
$$[\![\phi \wedge \psi]\!] = [\![\phi]\!] \cdot [\![\psi]\!],$$
$$[\![\phi \vee \psi]\!] = [\![\phi]\!] + [\![\psi]\!],$$
$$[\![\phi \to \psi]\!] = [\![\phi]\!] \Rightarrow [\![\psi]\!],$$
$$[\![\forall x \phi]\!] = \prod \{[\![\phi(x)]\!] : x \in \mathcal{M}^B\},$$
$$[\![\exists x \phi]\!] = \sum \{[\![\phi(x)]\!] : x \in \mathcal{M}^B\}.$$

The following two lemmas, which list the basic properties of Boolean values, can be proved by direct computation, using induction on $(\rho(x), \rho(y))$. The reader can either believe them or prove them himself (herself) or turn to the literature.

LEMMA 5.1

(i). $[\![x = x]\!] = 1$.
(ii). $[\![x(y)]\!] \leqslant [\![y \in x]\!]$.
(iii). $[\![x = y]\!] = [\![y = x]\!]$.
(iv). $[\![x = y]\!] \cdot [\![y = z]\!] \leqslant [\![x = z]\!]$.
(v). $[\![x = x_1]\!] \cdot [\![x \in y]\!] \leqslant [\![x_1 \in y]\!]$.
(vi). $[\![x = x_1]\!] \cdot [\![y \in x]\!] \leqslant [\![y \in x_1]\!]$.

LEMMA 5.2

(i). $[\![x = y]\!] \cdot [\![\phi(x)]\!] \leqslant [\![\phi(y)]\!]$.
(ii). $[\![(\exists y \in x)\phi(y)]\!] = \sum \{(x(y) \cdot [\![\phi(y)]\!]) : y \in \mathrm{dom}\,(x)\}$.
(iii). $[\![(\forall y \in x)\phi(y)]\!] = \prod \{(x(y) \Rightarrow [\![\phi(y)]\!]) : y \in \mathrm{dom}\,(x)\}$.

We shall now present the fundamental theorem of Boolean-valued models. Let $\phi(x_1, \ldots, x_n)$ be a formula, with variables in $\mathcal{M}^B$. We say that $\phi(x_1, \ldots, x_n)$ is *valid in* $\mathcal{M}^B$ if $[\![\phi(x_1, \ldots, x_n)]\!] = 1$.

THEOREM 5.3

(i). *Every axiom of the predicate logic is valid in* $\mathcal{M}^B$. *The rules of inference of the predicate logic if applied to formulas valid in* $\mathcal{M}^B$ *result in formulas valid in* $\mathcal{M}^B$.

(ii). *Every axiom of* ZF+AC *is valid in* $\mathcal{M}^B$. *Consequently, every statement provable in* ZF+AC *is valid in the Boolean-valued model.*

The proof of Theorem 5.3 (i) is by direct computation. The proof of (ii) is nontrivial, but since it is available in the literature and its knowledge is not essential for the practical applications of the method, we take the liberty of omitting it.

5.1.2. *Generic extensions.*

Now having introduced the Boolean-valued models, we shall step outside the model $\mathcal{M}$ and try to extend $\mathcal{M}$ to a larger model. Let $B \in \mathcal{M}$ be a complete Boolean algebra in $\mathcal{M}$. A set $G \subseteq B$ is called an *$\mathcal{M}$-generic ultrafilter* on B if

(5.1) G has the properties of an ultrafilter,

(5.2) $A \subseteq G$ and $A \in \mathcal{M}$ implies $\prod \{a : a \in A\} \in G$.

(Notice that G need not be in $\mathcal{M}$.)

We give another formulation of genericity: A subset $A \subseteq B$ is called a *partition* of B if $\sum \{a : a \in A\} = 1$ and if $u \cdot v = 0$ for all $u \neq v$ in A. Then condition (5.2) can be replaced by

(5.3) If $A \in \mathcal{M}$ is a partition of B, then there is a (unique) $u \in A$ such that $u \in G$.

Let G be an $\mathcal{M}$-generic ultrafilter on B. We define (by recursion on $\rho(x)$) the *interpretation* i_G of $\mathcal{M}^B$ by G:

(a) $i(0) = 0$,

(b) $i(x) = \{i(y) : x(y) \in G\}$.

The *generic extension* of $\mathcal{M}$ by G is the range of i_G:

$$\mathcal{M}[G] = \{i(x) : x \in \mathcal{M}^B\}.$$

We shall show that $\mathcal{M}[G]$ is a model which extends $\mathcal{M}$ and can be considered as an adjunction of G to $\mathcal{M}$.

LEMMA 5.4. *For each $x \in \mathcal{M}$, $i(\check{x}) = x$; hence $\mathcal{M} \subseteq \mathcal{M}[G]$.*

PROOF. By the $\in$-induction,

$$i(\check{0}) = 0,$$

$$i(\check{x}) = \{i(\check{y}) : \check{x}(\check{y}) \in G\} = x.$$

LEMMA 5.5. $G \in \mathcal{M}[G]$.

PROOF. We define the *canonical* generic ultrafilter $\underline{G} \in \mathcal{M}^B$ as follows:

$$\operatorname{dom}(\underline{G}) = \{\check{u} : u \in B\},$$

$$\underline{G}(\check{u}) = u \text{ for every } u \in B.$$

Obviously,

$$i(\underline{G}) = \{i(x) : \underline{G}(x) \in G\} = \{i(u) : u \in G\} = G.$$

If $x \in \mathcal{M}[G]$ and if $\underset{\sim}{x} \in \mathcal{M}_B$ is such that $i(\underset{\sim}{x}) = x$, then we say that $\underset{\sim}{x}$ is a *name* for x.

LEMMA 5.6. *If $\underset{\sim}{x}$, $\underset{\sim}{y}$ are names for x, y, then*

$$x \in y \leftrightarrow [\![\underset{\sim}{x} \in \underset{\sim}{y}]\!] \in G,$$
$$x = y \leftrightarrow [\![\underset{\sim}{x} = \underset{\sim}{y}]\!] \in G.$$

PROOF. By induction on $(\rho(x), \rho(y))$. We prove that $[\![\underset{\sim}{x} \in \underset{\sim}{y}]\!] \in G$ implies $x \in y$ and leave the rest to the reader. If $[\![\underset{\sim}{x} \in \underset{\sim}{y}]\!] \in G$, then

$$\sum_{z \in \mathrm{dom}(\underset{\sim}{y})} (\underset{\sim}{y}(z) \cdot [\![z = \underset{\sim}{x}]\!]) \in G.$$

By the genericity, there is $z \in \mathrm{dom}(\underset{\sim}{y})$ such that $\underset{\sim}{y}(z) \cdot [\![z = \underset{\sim}{x}]\!] \in G$; i.e., $\underset{\sim}{y}(z) \in G$ and $[\![z = \underset{\sim}{x}]\!] \in G$. Hence $i(z) \in y$, and by the induction hypothesis, $i(z) = x$; thus $x \in y$.

Now we present the main theorem of generic models. It expresses the connection between Boolean values and the truth in generic extension.

THEOREM 5.7. *Let ϕ be a formula. If $\underset{\sim}{x}_1, \dots, \underset{\sim}{x}_n$ are names for $x_1, \dots, x_n \in \mathcal{M}[G]$, then*

$$\mathcal{M}[G] \models \phi(x_1, \dots, x_n) \text{ if and only if } [\![\phi(\underset{\sim}{x}_1, \dots, \underset{\sim}{x}_n)]\!] \in G.$$

PROOF. By induction on the complexity of ϕ. Use Lemma 5.6 and the genericity of G. E.g.,

$$\mathcal{M}[G] \models \neg\phi \leftrightarrow \neg(\mathcal{M}[G] \models \phi) \leftrightarrow \neg([\![\phi]\!] \in G) \leftrightarrow -[\![\phi]\!] \in G \leftrightarrow [\![\neg\phi]\!] \in G.$$

COROLLARY 5.8.

(i). *$\mathcal{M}[G]$ is a model of* ZF+AC.

(ii). *$\mathcal{M}[G]$ is the least model $\mathcal{N}$ of* ZF *such that $\mathcal{M} \subseteq \mathcal{N}$ and $G \in \mathcal{N}$.*

PROOF. We proved $\mathcal{M} \subseteq \mathcal{M}[G]$ and $G \in \mathcal{M}(G)$. If $\mathcal{N} \supseteq \mathcal{M}$ is a model of ZF and if $G \in \mathcal{N}$, then for all $\alpha \in \mathcal{M}$, $\mathcal{M}^B_\alpha \in \mathcal{N}$ and $i_G|\mathcal{M}^B_\alpha \in \mathcal{N}$ (the definition of $i_G|\mathcal{M}^B_\alpha$ is absolute for every $\mathcal{N}$ containing G and $\mathcal{M}^B_\alpha$). Thus $\mathcal{M}[G] \subseteq \mathcal{N}$.

Until now, we carefully avoided the question of whether a generic ultrafilter exists. As a matter of fact, if the algebra B is atomless, then G cannot be an element of $\mathcal{M}$; thus we cannot prove that G exists. Still, we shall always make the following assumption:

In each case, we shall assume that a generic ultrafilter exists.

The reasons are practical. We use generic extensions to establish consistency results; we could do as well without generic ultrafilters and work entirely with Boolean-valued models. For instance, we shall present a generic model $\mathscr{M}[G]$, in which there is a nonconstructible set. Instead, we could prove that in the corresponding $\mathscr{M}^B$, the value $[\![$every set is constructible$]\!]$ is not 1. In each case, we conclude that the Axiom of Constructibility is unprovable in ZF.

(We actually have a justification to assume that a generic ultrafilter exists. In the Boolean-valued model, it is valid that the canonical generic ultrafilter $\underset{\sim}{G}$ is generic; see Problems 1 and 2.)

5.1.3. *Forcing.*

In practical applications of generic models, one does not work much with Boolean values. Instead, one uses so-called *forcing conditions.*

Let $\mathscr{M}$ be a fixed transitive model of ZF+AC. Let $(P, \leqslant)$ be a partially ordered set in $\mathscr{M}$. The elements of P are called *forcing conditions*; p is *stronger* than q if $p \leqslant q$. Two conditions p and q are *compatible* if there is a condition r stronger than both p and q. A set $D \subseteq P$ is said to be *dense* in P if for each $p \in P$ there is $d \in D$ stronger than p.

Let G be a set of conditions, not necessarily in $\mathscr{M}$. We say that G is *$\mathscr{M}$-generic*, if:

(5.4) (a) $x \in G$ and $x \leqslant y$ implies $y \in G$;
(b) the elements of G are pairwise compatible;
(c) if D is dense in P and $D \in \mathscr{M}$, then $D \cap G \neq 0$.

Note that if B is a complete Boolean algebra in $\mathscr{M}$ and if we let P be the set of all nonzero elements of B, then $G \subseteq P$ is generic if and only if it is a generic ultrafilter. Moreover, if B is a complete Boolean algebra in $\mathscr{M}$, and P is dense in $B - \{0\}$, then:

(i) if G is a generic ultrafilter on B, then $G_1 = G \cap P$ satisfies (5.4);
(ii) if $G_1 \subseteq P$ satisfies (5.4), then $G = \{u \in B : (\exists p \in P)\, p \leqslant u\}$ is a generic ultrafilter.

(We leave the easy proofs to the reader.)

Thus G_1 is definable in terms of G and elements of $\mathscr{M}$, and vice versa, which we may formulate as follows:

(5.5) $$\mathscr{M}[G_1] = \mathscr{M}[G]$$

(the model $\mathscr{M}[G]$ is the least model $\mathscr{N}$ such that $\mathscr{N} \supseteq \mathscr{M}$ and $G_1 \in \mathscr{N}$). Hence if $(P, <)$ is a dense subset of a complete Boolean algebra, a generic set of conditions determines a generic extension of $\mathscr{M}$. However, every partially ordered set can be associated with a complete Boolean algebra B (which is unique up to isomorphism) such that we get the correspondence (5.5) between generic subsets of P and generic ultrafilters on B. We shall sketch the proof of this fact.

Consider a topological space. A *regular open* set is a set which is equal to the interior of its closure. It is a well-known fact that the family of all regular open sets of a given space is a complete Boolean algebra (with operations $\cup$ and $\cap$ and ordering $\subseteq$; the Boolean complement is *not* the set theoretical complement).

Let $(P, \leqslant)$ be a partially ordered set. Let us endow P with the topology generated by the basic open sets $[p] = \{q \in P : q \leqslant p\}$. Let us denote by RO$(P)$ the complete Boolean algebra of all regular open sets in this space. Let e be the following mapping of P into RO(P):

$$e(P) = \text{the interior of the closure of } [p].$$

One can verify that e has the following properties:

(5.6) (i) if $p \leqslant q$, then $e(P) \subseteq e(Q)$;
(ii) p, q are compatible if and only if $e(P) \cap e(Q) \neq 0$;
(iii) the set $\{e(p) : p \in P\}$ is dense in RO$(P) - \{0\}$.

We call e the *natural homomorphism* of P into RO(P). (The reader will find an alternate construction of RO(P) in Problems 3 and 4.)

Now let $(P, \leqslant)$ be a partially ordered set in $\mathcal{M}$, and let e be the natural homomorphism of P into RO(P) (in $\mathcal{M}$). We ask the reader to use (5.6) to verify the following correspondence between generic subsets of P and generic ultrafilters on RO(P):

(i) if G is a generic ultrafilter on B, then $G_1 = e_{-1}(G)$ satisfies (5.4);
(ii) if $G_1 \subseteq P$ satisfies (5.4), then $G = \{u \in B : (\exists p \in G_1)[e(p) \leqslant u]\}$ is a generic ultrafilter.

As a consequence, we have $\mathcal{M}[G_1] = \mathcal{M}[G]$.

Thus every partially ordered set in $\mathcal{M}$ determines a generic extension $\mathcal{M}[G]$. The natural embedding e gives us, however, more information and enables us to reformulate the main theorem of generic models. (We shall consider P a subset of RO(P) and write p instead of $e(p)$.)

Let $(P, \leqslant) \in \mathcal{M}$, let $B = \text{RO}(P)$ and let $e : P \to B$ be the natural embedding (in $\mathcal{M}$). If $\phi(\underline{x}_1, \ldots, \underline{x}_n)$ is a formula with variables in $\mathcal{M}^B$, then we say that a condition p *forces* $\phi(\underline{x}_1, \ldots, \underline{x}_n)$,

$$p \Vdash \phi(\underline{x}_1, \ldots, \underline{x}_n),$$

if $e(p) \leqslant [\![\phi(\underline{x}_1, ..., \underline{x}_n)]\!]$. The *forcing relation* $\Vdash$ has the following properties, which can be derived from the properties of Boolean values:

$$\begin{array}{lll} p \Vdash \neg\phi & \text{iff} & \text{no } q \leqslant p \text{ forces } \phi, \\ p \Vdash \phi \wedge \psi & \text{iff} & p \Vdash \phi \text{ and } p \Vdash \psi, \\ p \Vdash \phi \vee \psi & \text{iff} & (\forall q \leqslant p)(\exists r \leqslant q)[r \Vdash \phi \vee r \Vdash \psi], \\ p \Vdash \forall x\phi & \text{iff} & (\forall \underline{x} \in \mathcal{M}^B)\, p \Vdash \phi(x), \\ p \Vdash \exists x\phi & \text{iff} & (\forall q \leqslant p)(\exists r \leqslant q)(\exists x \in \mathcal{M}^B)\, r \Vdash \phi(x); \end{array}$$

and, if we say that p *decides* ϕ just in case either $p \Vdash \phi$ or $p \Vdash \neg\phi$, then

$$\forall p(\exists q \leqslant p)\; q \text{ decides } \phi.$$

The main theorem of generic models now takes the following form:

THEOREM 5.9. *Let ϕ be a formula. If $G \subseteq P$ is $\mathcal{M}$-generic and if $\underline{x}_1, ..., \underline{x}_n$ are names for $x_1, ..., x_n \in \mathcal{M}[G]$, then*

$$\mathcal{M}[G] \vDash \phi(x_1, ..., x_n) \textit{ if and only if } (\exists p \in G) p \Vdash \phi(\underline{x}_1, ..., \underline{x}_n).$$

5.1.4. *Examples of generic models.*

EXAMPLE 5.10. Let $\mathcal{M}$ be a transitive model of ZF+AC. We shall describe a set of conditions which adds a new set of integers to the ground model. Let $(P, <)$ be the set of all finite sequences of 0's and 1's, partially ordered by $\supseteq$ (i.e., p is stronger than q if $p \supseteq q$). Let G be an $\mathcal{M}$-generic set of conditions. It is easy to see that for each n, the set $\{p \in P : p(n) = 1\}$ is dense in P, so that there is $p \in G$ such that $n \in \operatorname{dom}(p)$. Also, if p and q are compatible, then either $p \supseteq q$ or $q \supseteq p$. Thus

(5.7) $\qquad g = \bigcup \{p : p \in G\}$ is a function from ω into $\{0, 1\}$.

Let $\underline{z}$ be the following name: $\operatorname{dom}(\underline{z}) = \{\check{n} : n \in \omega\}$ and $\underline{z}(\check{n}) = \sum \{p \in P : p(\check{n}) = 1\}$; $\underline{z}$ is a name for the set $z = \{n : g(n) = 1\}$. We shall prove

(5.8) $\qquad G \notin \mathcal{M},$

which in turn gives:

(5.9) $\qquad$ there is a set $z \subseteq \omega$ such that $z \in \mathcal{M}[G]$ and $z \notin \mathcal{M}$.

Assume that $G \in \mathcal{M}$. For each $q \in P$, either $q^\frown 0$ or $q^\frown 1$ is not in G. Hence $\{p : p \notin G\}$ is dense in P. This contradicts the genericity of G.

EXAMPLE 5.11. We shall describe a set of conditions which adds a new subset of ω_1 to the ground model without adding new subsets of ω. Let

$(P, \leqslant)$ be (in $\mathcal{M}$) the set of all α-sequences of 0's and 1's, for all $\alpha < \omega_1$, and order P by $\supseteq$. Let G be an $\mathcal{M}$-generic set of conditions. As in Example 1, we get $G \notin \mathcal{M}$ and G determines a subset of ω_1 which is not in $\mathcal{M}$. However, we have to show that

(i) every subset of ω in $\mathcal{M}[G]$ is in $\mathcal{M}$,

(ii) $(\aleph_1)^{\mathcal{M}[G]} = (\aleph_1)^{\mathcal{M}}$.

Both (i) and (ii) follow from the following:

(5.10) If $f \in \mathcal{M}[G]$ is a function from ω to $\mathcal{M}$, then $f \in \mathcal{M}$.

The property (5.10) follows from specific properties of $(P, \leqslant)$. We prove a general lemma which is one of the most typical examples of using properties of $(P, \leqslant)$ to establish properties of $\mathcal{M}[G]$, and leave the reader to verify that the provisions of the lemma apply in our special case.

LEMMA 5.12. *Assume that (in $\mathcal{M}$) for every descending sequence*

$$p_0 \geqslant p_1 \geqslant \dots \geqslant p_n \geqslant \dots \qquad (n \in \omega)$$

of conditions there exists $p \in P$ such that $p \leqslant p_n$ for all n. Then $\mathcal{M}[G]$ has the property (5.10).

PROOF. A subset $D \subseteq P$ is *open dense* if it is dense and if $p_1 \leqslant p_2$ and $p_2 \in D$ implies $p_1 \in D$. First we show that under the assumption of the lemma, any intersection of countably many open dense sets is open dense. Let D_n, $n \in \omega$, be open dense. If $p \in P$, we let, by recursion, p_n be an element of D_n stronger than all p_m, $m < n$; then we let q be stronger than all p_n. Clearly, $q \leqslant p$ and $q \in \bigcap_{n=0}^{\infty} D_n$.

Let $f : \omega \to \mathcal{M}$ and $f \in \mathcal{M}[G]$, and let $\underset{\sim}{f}$ be a name for f. Let $A \in \mathcal{M}$ be such that $\operatorname{rng}(f) \subseteq A$ and assume that

$$u = [\underset{\sim}{f} \text{ is a mapping of } \check{\omega} \text{ into } \check{A}] = 1;$$

we can assume this since $u \in G$, and we can restrict ourselves to $P' = \{p \in P : p \leqslant u\}$. For each n, the set

$$D_n = \{p : (\exists x \in A) p \Vdash \underset{\sim}{f}(\check{n}) = \check{x}\}$$

is open dense. Hence $D = \bigcap_{n=0}^{\infty} D_n$ is dense and so there exists $p \in D$ such that $p \in G$. For each n, there is x_n such that $p \Vdash \underset{\sim}{f}(\check{n}) = \check{x}_n$. If we denote by g the function $g : n \mapsto x_n$ (obviously, $g \in \mathcal{M}$), we have $p \Vdash \underset{\sim}{f} = \check{g}$ and therefore $f = g$.

In a similar fashion, if $\aleph_\alpha$ is a regular cardinal in $\mathcal{M}$, and if we take as forcing conditions transfinite sequences of 0's and 1's of length less than

ω_α, we get a generic model with a new subset of ω_α, without adding subsets of smaller cardinals. (For the analogue of Lemma 5.1.6, see Problem 5.)

5.2. Symmetric submodels of generic models

The generic extensions satisfy all axioms of set theory including the Axiom of Choice. Still they can be used to establish the independence of the Axiom of Choice, by considering certain submodels of generic models. The idea behind *symmetric models* is analogous to the ideas used in connection with permutation models. Although the universe of ZF does not admit $\in$-automorphisms, the Boolean-valued universe does (via automorphisms of the Boolean algebra), and the automorphisms of the *names* of elements of $\mathcal{M}[G]$ can be used to construct submodels of $\mathcal{M}[G]$ resembling the permutation models.

Let B be a Boolean algebra. An *automorphism* π of B is a one-to-one mapping of B onto itself which preserves the Boolean operations: $\pi(u+v) = \pi(u)+\pi(v)$, $\pi(-u) = -\pi(u)$ etc. If B is a complete Boolean algebra and π is an automorphism of B, then π preserves infinite sums and products as well.

Let $\mathcal{M}$ be a transitive model of ZF+AC, let B be a complete Boolean algebra in $\mathcal{M}$, and let $\mathcal{M}^B$ be the corresponding Boolean-valued model. (Needless to say, until the appearance of a generic ultrafilter we stay within $\mathcal{M}$.) Let π be an automorphism of B. We can extend π to $\mathcal{M}^B$ as follows, by recursion on $\rho(x)$:

(i) $\pi 0 = 0$;

(ii) assuming that π has been defined for all $y \in \mathrm{dom}\,(x)$, let $\mathrm{dom}\,(\pi x) = \pi''\,\mathrm{dom}\,(x)$, and $(\pi x)(\pi y) = \pi(x(y))$ for all $\pi y \in \mathrm{dom}\,(\pi x)$.

Clearly, π is a one-to-one function of $\mathcal{M}^B$ onto itself, and $\pi\check{x} = \check{x}$ for every $x \in \mathcal{M}$.

LEMMA 5.13. *Let $\phi(x_1, ..., x_n)$ be a formula with variables in $\mathcal{M}^B$. Then*

$$[\![\phi(\pi x_1, ..., \pi x_n)]\!] = \pi[\![\phi(x_1, ..., x_n)]\!].$$

PROOF. First, one proves the lemma for $x \in y$ and $x = y$ by induction on $(\rho(x), \rho(y))$. The rest follows by induction on the complexity of ϕ.

Let $\mathscr{G}$ be a group of automorphisms of B. For each $x \in \mathcal{M}^B$, let

$$\mathrm{sym}_{\mathscr{G}}\,(x) = \{\pi \in \mathscr{G} : \pi(x) = x\};$$

$\mathrm{sym}_{\mathscr{G}}\,(x)$ is a subgroup of $\mathscr{G}$. A nonempty set $\mathscr{F}$ of subgroups of $\mathscr{G}$ is called a *normal filter* on $\mathscr{G}$ iff for all subgroups H, K of $\mathscr{G}$,

(i) if $K \in \mathscr{F}$ and $H \supseteq K$, then $H \in \mathscr{F}$;
(ii) if $H \in \mathscr{F}$ and $K \in \mathscr{F}$, then $H \cap K \in \mathscr{F}$;
(iii) if $\pi \in \mathscr{G}$ and $H \in \mathscr{F}$, then $\pi H \pi^{-1} \in \mathscr{F}$.

Let $\mathscr{F}$ be fixed. We say that $x \in \mathscr{M}^B$ is *symmetric* if $\operatorname{sym}(x) \in \mathscr{F}$. The class $\mathrm{HS} \subseteq \mathscr{M}^B$ of all *hereditarily symmetric* names is defined by recursion:
(a) $0 \in \mathrm{HS}$;
(b) if $\operatorname{dom}(x) \subseteq \mathrm{HS}$ and if x is symmetric, then $x \in \mathrm{HS}$.

Since $\pi(\check{x}) = \check{x}$ for each $\pi \in \mathscr{G}$ and each $x \in \mathscr{M}$, it follows that HS contains each $\check{x}$.

Now let G be an $\mathscr{M}$-generic ultrafilter on B. Let i_G be the interpretation of $\mathscr{M}^B$ by G. We let

$$\mathscr{N} = \{i_G(x) : x \in \mathrm{HS}\},$$

we have $\mathscr{M} \subseteq \mathscr{N} \subseteq \mathscr{M}[G]$.

THEOREM 5.14. *$\mathscr{N}$ is a model of* ZF.

PROOF. Assume that $\mathscr{M}$ is a proper class and we show that it is transitive, closed under Gödel operations and almost universal. If $\mathscr{M}$ is a set, then the proof can be modified to show that $\mathscr{N}$ is a model of ZF.

The transitivity of $\mathscr{N}$ follows from the fact that if $x \in \mathrm{HS}$, then $\operatorname{dom}(x) \subseteq \mathrm{HS}$.

For all $x, y \in \mathscr{M}^B$, one can define canonically $z_1, \ldots, z_8 \in \mathscr{M}^B$ such that $[\![z_i = \mathscr{F}_i(x, y)]\!] = 1$ for $i = 1, \ldots, 8$ and $\operatorname{sym}(z_i) \supseteq \operatorname{sym}(x) \cap \operatorname{sym}(y)$ and $z_i \in \mathrm{HS}$ whenever $x, y \in \mathrm{HS}$; we leave the details to the reader. It follows that $\mathscr{N}$ is closed under Gödel operations $\mathscr{F}_1, \ldots, \mathscr{F}_8$.

To show that $\mathscr{N}$ is almost universal, notice that if X is a subset of $\mathscr{N}$, then $X \subseteq i_G''(\mathrm{HS} \cap \mathscr{M}^B_\alpha)$ for some α. Thus it suffices to show that each $Y = i_G''(\mathrm{HS} \cap \mathscr{M}^B_\alpha)$ is in $\mathscr{N}$. Let $\underline{Y} \in \mathscr{M}^B$ be as follows: $\operatorname{dom}(\underline{Y}) = \mathrm{HS} \cap \mathscr{M}^B_\alpha$ and $\underline{Y}(x) = 1$ for each $x \in \operatorname{dom}(\underline{Y})$. $\underline{Y}$ is a name for Y, and $\operatorname{dom}(\underline{Y}) \subseteq \mathrm{HS}$; hence we only have to show that $\underline{Y}$ is symmetric. We will actually show that $\operatorname{sym}(\underline{Y}) = \mathscr{G}$. If $x \in \mathscr{M}^B$ and $\pi \in \mathscr{G}$, then $\operatorname{sym}(\pi x) = \pi \cdot \operatorname{sym}(x) \cdot \pi^{-1}$; hence if $x \in \mathrm{HS}$, then $\pi x \in \mathrm{HS}$ for each $\pi \in \mathscr{G}$. Also, if $x \in \mathscr{M}^B_\alpha$, then $\pi x \in \mathscr{M}^B_\alpha$ for each $\pi \in \mathscr{G}$. Hence $\pi''(\mathrm{HS} \cap \mathscr{M}^B_\alpha) = \mathrm{HS} \cap \mathscr{M}^B_\alpha$, and so $\pi(\underline{Y}) = \underline{Y}$, for all $\pi \in \mathscr{G}$.

We call $\mathscr{N}$ a *symmetric extension* of $\mathscr{M}$. In the subsequent sections we use symmetric extensions to establish the independence of the Axiom of Choice.

5.3. The basic Cohen model

Let $\mathscr{M}$ be a transitive model of ZF+AC. Consider the following set $(P, <)$ of forcing conditions: the elements of P are finite functions p with values 0, 1, and $\operatorname{dom}(p) \subseteq \omega \times \omega$; the ordering is by $\supseteq$. Let $B = \operatorname{RO}(P)$ in $\mathscr{M}$.

We shall define $\mathscr{G}$ and $\mathscr{F}$ in such a way that the Axiom of Choice will fail in the symmetric extension $\mathscr{N}$. Before doing that, we shall describe the motivation. Let G be an $\mathscr{M}$-generic ultrafilter on B. For each $n \in \omega$, let x_n be the following real number (a subset of ω):

$$x_n = \{m \in \omega : (\exists p \in G)\ p(n, m) = 1\}. \tag{5.11}$$

We intend to construct $\mathscr{N}$ in such a way that the set

$$A = \{x_n : n \in \omega\}$$

is in $\mathscr{N}$ but the enumeration $n \mapsto x_n$ is not; we want that

$$\mathscr{N} \models A \text{ cannot be well-ordered.}$$

The reals x_n, $n \in \omega$, have canonical names $\underset{\sim}{x}_n$:

$$\underset{\sim}{x}_n(\check{m}) = u_{nm} = \sum \{p \in P : p(n, m) = 1\} \quad \text{for all } m, n \in \omega; \tag{5.12}$$

so does the set A:

$$\operatorname{dom}(\underset{\sim}{A}) = \{\underset{\sim}{x}_n : n \in \omega\}, \qquad \underset{\sim}{A}(\underset{\sim}{x}_n) = 1 \quad \text{for all } n \in \omega. \tag{5.13}$$

Let π be a permutation of ω; π induces an order-preserving 1-1 mapping of $(P, \leqslant)$ onto itself as follows:

$$\operatorname{dom}(\pi p) = \{(\pi n, m) : (n, m) \in \operatorname{dom}(p)\},$$
$$(\pi p)(\pi n, m) = p(n, m).$$

In turn, this π induces an automorphism of B:

$$\pi u = \sum \{\pi p : p \leqslant u\}.$$

Let $\mathscr{G}$ be the group of all automorphisms of B which are induced by permutations of ω, as described above. For every finite $e \subseteq \omega$, let

$$\operatorname{fix}(e) = \{\pi \in \mathscr{G} : \pi n = n \text{ for each } n \in e\}.$$

Let $\mathscr{F}$ be the filter on $\mathscr{G}$ generated by $\operatorname{fix}(e)$, e finite; i.e.,

$$H \in \mathscr{F} \leftrightarrow \exists e\ (H \supseteq \operatorname{fix}(e)).$$

We leave it to the reader to verify that $\mathscr{F}$ is a normal filter on $\mathscr{G}$.

$\mathscr{G}$ and $\mathscr{F}$ determine the class HS of hereditarily symmetric names, and if we let G be an $\mathscr{M}$-generic ultrafilter on B, the interpretation i_G of HS yields a symmetric model $\mathscr{N}$ of ZF.

LEMMA 5.15. *In the model $\mathscr{N}$, the set of all real numbers cannot be well-ordered.*

PROOF. We show that the set $A = \{x_n : n \in \omega\}$, where x_n are the reals defined in (5.11), cannot be well-ordered in $\mathscr{N}$. To start with, all the reals x_n, as well as the set A, are in $\mathscr{N}$, since their names, defined in (5.12) and (5.13) are symmetric: For all $\pi \in \mathscr{G}$ and all n, m, we have $\pi(u_{nm}) = u_{\pi n, m}$; hence $\pi(\underset{\sim}{x}_n) = \underset{\sim}{x}_{\pi n}$, and consequently, $\operatorname{sym}(\underset{\sim}{x}_n) = \operatorname{fix}\{n\} \in \mathscr{F}$. Also $\pi(\underset{\sim}{A}) = \underset{\sim}{A}$ for all $\pi \in \mathscr{G}$, and so $A \in \mathscr{N}$.

Next, notice that these reals are pairwise distinct: we show that $[\![\underset{\sim}{x}_i = \underset{\sim}{x}_j]\!] = 0$ whenever $i \neq j$. For assume that there is $p \in P$ such that $p \Vdash \underset{\sim}{x}_i = \underset{\sim}{x}_j$. There is m such that neither (i, m) nor (j, m) belong to dom (p); let $q \supseteq p$ be such that $q(i, m) = 1$ and $q(j, m) = 0$. Then

$$q \Vdash \check{m} \in \underset{\sim}{x}_i \quad \text{and} \quad q \Vdash \check{m} \notin \underset{\sim}{x}_j,$$

hence $q \Vdash \underset{\sim}{x}_i \neq \underset{\sim}{x}_j$ although $q \leqslant p$, a contradiction.

We shall show that in $\mathscr{N}$ there is no one-to-one mapping of ω onto A. Assume that there is some, and let $\underset{\sim}{f} \in HS$ be its name. Thus for some $p_0 \in G$,

$$p_0 \Vdash \underset{\sim}{f} \text{ is a one-to-one mapping of } \check{\omega} \text{ onto } \underset{\sim}{A}.$$

We shall find $q \leqslant p_0$ such that

$$q \Vdash \underset{\sim}{f} \text{ is not a function,} \tag{5.14}$$

which will be a contradiction.

Let e be a finite subset of ω such that $\operatorname{sym}(\underset{\sim}{f}) \supseteq \operatorname{fix}(e)$. There exist $i \in \omega$, $p \leqslant p_0$ and $n \notin e$ such that

$$p \Vdash \underset{\sim}{f}(\check{i}) = \underset{\sim}{x}_n.$$

We shall find $\pi \in \mathscr{G}$ such that:

(5.15) (i) πp and p are compatible;
(ii) $\pi \in \operatorname{fix}(e)$;
(iii) $\pi n \neq n$.

Then we will have $\pi \underline{f} = \underline{f}$ by (ii), $\pi \check{i} = \check{i}$, and since

$$\pi p \Vdash (\pi \underline{f})(\pi \check{i}) = \pi x_n,$$

we will have

$$q = p \cup \pi p \Vdash (\underline{f}(\check{i}) = \underline{x}_n \wedge \underline{f}(\check{i}) = \underline{x}_{\pi n}),$$

and since $[\![\underline{x}_n = \underline{x}_{\pi n}]\!] = 0$ by (iii), we will have (5.14).

To get π, let n' be such that $n' \notin e$ and $(n', m) \notin \operatorname{dom}(p)$ for any m. Let π be the permutation of ω which interchanges n and n', and $\pi(i) = i$ for all $i \neq n, n'$. This permutation satisfies (5.15) and the proof is complete.

As a consequence, we have:

THEOREM 5.16. *The Axiom of Choice is unprovable in set theory.*

5.4. The second Cohen model

Let $\mathcal{M}$ be a transitive model of ZF+AC. We shall construct a symmetric extension $\mathcal{N}$ of $\mathcal{M}$ in which there is a countable family of pairs which has no choice function.

In the symmetric model of Section 5.3, the non-well-orderable set consisted of real numbers. Obviously, we cannot use real numbers in the present case to construct the pairs, since pairs of reals do have a choice function. However, the method will work when the intended elements of the pairs are sets of reals.

Let $(P, \leqslant)$ be the set of all finite functions p with values 0, 1 and $\operatorname{dom}(p) \subseteq (\omega \times \{0, 1\} \times \omega) \times \omega$; let $\leqslant\ =\ \supseteq$. Let $B = \operatorname{RO}(P)$ in $\mathcal{M}$.

Let G be an $\mathcal{M}$-generic ultrafilter on B. We define the following elements of $\mathcal{M}[G]$, together with their canonical names:

$$\begin{aligned}
x_{n\varepsilon i} &= \{j \in \omega : (\exists p \in G)[p(n\varepsilon i, j) = 1]\},\\
\underline{x}_{n\varepsilon i}(\check{j}) &= u_{n\varepsilon i, j} = \textstyle\sum \{p \in P : p(n\varepsilon i, j) = 1\} \quad (n, i, j \in \omega,\ \varepsilon = 0, 1),\\
X_{n\varepsilon} &= \{x_{n\varepsilon i} : i \in \omega\},\\
\operatorname{dom}(\underline{X}_{n\varepsilon}) &= \{\underline{x}_{n\varepsilon} : i \in \omega\}, \qquad \underline{X}_n(\underline{x}_{n\varepsilon i}) = 1 \quad \text{for all } i \in \omega,\\
P_n &= \{X_{n0}, X_{n1}\},\\
\operatorname{dom}(\underline{P}_n) &= \{\underline{X}_{n0}, \underline{X}_{n1}\}, \qquad \underline{P}_n(\underline{X}_{n\varepsilon}) = 1 \quad (\varepsilon = 0, 1),\\
A &= \{P_n : n \in \omega\},\\
\operatorname{dom}(\underline{A}) &= \{\underline{P}_n : n \in \omega\}, \qquad \underline{A}(\underline{P}_n) = 1 \quad \text{for all } n \in \omega.
\end{aligned}$$

Each $x_{n\varepsilon i}$ is a real (a subset of ω), $X_{n\varepsilon}$ are sets of reals. Similarly as in Lemma 5.15, we have

$$[\![x_{n\varepsilon i} = x_{n'\varepsilon' i'}]\!] = 0$$

whenever $(n, \varepsilon, i) \neq (n', \varepsilon', i')$.

Every permutation π of $(\omega \times \{0, 1\} \times \omega)$ induces a one-to-one preserving mapping π of P onto itself:

$$(\pi p)(\pi(n\varepsilon i), j) = p(n\varepsilon i, j),$$

and, in turn, an automorphism of B. Clearly,

$$\pi(u_{n\varepsilon i, j}) = u_{\pi(n\varepsilon i), j}$$

and

$$\pi(\underline{x}_{n\varepsilon i}) = \underline{x}_{\pi(n\varepsilon i)}.$$

We intend to define the group $\mathcal{G}$ and the filter $\mathcal{F}$ in such a way that the names $\underline{x}_{n\varepsilon i}$, $\underline{X}_{n\varepsilon}$, $\underline{P}_n$ and $\underline{A}$ are all symmetric. Let $\mathcal{G}$ be the group of all automorphisms of B induced by those permutations π of $\omega \times \{0, 1\} \times \omega$ which satisfy the following conditions: if $\pi(n\varepsilon i) = (\bar{n}\bar{\varepsilon}\bar{\imath})$, then

(i) $\bar{n} = n$;

(ii) for each n, either $\forall i\ (\bar{\varepsilon} = \varepsilon)$ or $\forall i\ (\bar{\varepsilon} \neq \varepsilon)$.

Notice that for each $\pi \in \mathcal{G}$,

$$\pi(\underline{X}_{n\varepsilon}) = \underline{X}_{n\varepsilon} \text{ or } \underline{X}_{n, 1-\varepsilon},$$
$$\pi(\underline{P}_n) = \underline{P}_n, \qquad \pi(\underline{A}) = \underline{A}.$$

For each finite $e \subseteq \omega \times \{0, 1\} \times \omega$, let

$$\text{fix}\,(e) = \{\pi \in \mathcal{G} : (\forall s \in e)[\pi s = s]\},$$

and let $\mathcal{F}$ be the filter on $\mathcal{G}$ generated by

$$\{\text{fix}\,(e) : e \subseteq \omega \times \{0, 1\} \times \omega,\ e \text{ finite}\}.$$

$\mathcal{F}$ is a normal filter. Let HS be the class of all hereditarily symmetric names in $\mathcal{M}^B$. Let $\mathcal{N}$ be the symmetric extension of $\mathcal{M}$, given by the interpretation of HS by G.

LEMMA 5.17. *The sets* $x_{n\varepsilon i}$, $X_{n\varepsilon}$, P_n *and* A *are in the model* $\mathcal{N}$ *for all* $n, i \in \omega$, $\varepsilon = 0, 1$.

PROOF. Note that sym $(\underline{x}_{n\varepsilon i})$ = fix $\{(m\varepsilon i)\}$, sym $(\underline{X}_{n\varepsilon})$ = fix $\{(n\varepsilon i)\}$ (where i is arbitrary), sym $(\underline{P}_n) = \mathcal{G}$ and sym $(\underline{A}) = \mathcal{G}$.

LEMMA 5.18. *The set A is countable in $\mathcal{N}$.*

PROOF. It suffices to find a symmetric name for the function $g : n \mapsto P_n$. Let $\underset{\sim}{g} \in \mathcal{M}^B$ be as follows: [2]

$$\operatorname{dom}(\underset{\sim}{g}) = \{(\check{n}, \underset{\sim}{P_n})^B : n \in \omega\}, \qquad \underset{\sim}{g}(x) = 1 \quad \text{for all } x \in \operatorname{dom}(\underset{\sim}{g}).$$

Since $\pi(\check{n}) = \check{n}$ and $\pi(\underset{\sim}{P_n}) = \underset{\sim}{P_n}$ for all $\pi \in \mathcal{G}$, $n \in \omega$, it follows that $\pi(\underset{\sim}{g}) = \underset{\sim}{g}$ for all $\pi \in \mathcal{G}$.

LEMMA 5.19. *There is no function $f \in \mathcal{N}$ such that* $\operatorname{dom}(f) = A$ *and* $f(P_n) \in P_n$ *for all n.*

PROOF. Assume that f is such a function; let $\underset{\sim}{f}$ be a symmetric name for f and let $p_0 \in G$ be such that

$$p_0 \Vdash \underset{\sim}{f} \text{ is a function on } \underset{\sim}{A} \text{ and } \forall n\, (\underset{\sim}{f}(\underset{\sim}{P_n}) \in \underset{\sim}{P_n}).$$

We shall find $q \leqslant p_0$ such that

$$q \Vdash \underset{\sim}{f} \text{ is not a function,} \tag{5.16}$$

which will be a contradiction.

Let e be a finite subset of $\omega \times \{0, 1\} \times \omega$ such that $\operatorname{sym}(\underset{\sim}{f}) \supseteq \operatorname{fix}(e)$. There exist $n \in \omega$, ε_0 (let us assume that $\varepsilon_0 = 0$) and $p \leqslant p_0$ such that

$$p \Vdash \underset{\sim}{f}(\underset{\sim}{P_n}) = \underset{\sim}{X}_{n0}.$$

We shall find $\pi \in \mathcal{G}$ such that

(5.17) (i) πp and p are compatible,
(ii) $\pi \in \operatorname{fix}(e)$,
(iii) $\pi(\underset{\sim}{X}_{n0}) = \underset{\sim}{X}_{n1}$.

Then we will have $\pi\underset{\sim}{f} = \underset{\sim}{f}$ by (ii), $\pi\underset{\sim}{P_n} = \underset{\sim}{P_n}$, and since

$$\pi p \Vdash (\pi\underset{\sim}{f})(\pi\underset{\sim}{P_n}) = \pi\underset{\sim}{X}_{n0},$$

we will have

$$q = p \cup \pi p \Vdash (\underset{\sim}{f}(\underset{\sim}{P_n}) = \underset{\sim}{X}_{n0} \wedge \underset{\sim}{f}(\underset{\sim}{P_n}) = \underset{\sim}{X}_{n1}),$$

and since $[\![\underset{\sim}{X}_{n0} = \underset{\sim}{X}_{n1}]\!] = 0$, we will have (5.16).

To get π, let k be such that

$$(\forall i \geqslant k)(\forall \varepsilon)[(n \varepsilon i) \notin \operatorname{dom}(p)].$$

[2] The Boolean pair $(x, y)^B$ for $x, y \in \mathcal{M}^B$ is defined as follows: $z = \{x, y\}^B$ has domain $\{x, y\}$ and $z(x) = z(y) = 1$; $(x, y)^B = \{\{x, x\}^B, \{x, y\}^B\}^B$. Obviously, any interpretation of $(x, y)^B$ gives a pair (ix, iy).

Let π be the permutation of $\omega \times \{0, 1\} \times \omega$ defined as follows:

$$\pi(n, 0, i) = \begin{cases} (n, 1, i+k) & \text{if } i < k, \\ (n, 1, i-k) & \text{if } k \leqslant i < 2k, \\ (n, 1, i) & \text{if } 2k < i, \end{cases}$$

$$\pi(n, 1, i) \ldots \text{accordingly}$$

$$\pi(n'\varepsilon i) = n'\varepsilon i \quad \text{for all } n' \neq n.$$

This permutation satisfies (5.17), and the proof is complete.

As a consequence, we have:

THEOREM 5.20. *The Axiom of Choice for countable families of pairs is unprovable in set theory.*

5.5. Independence of the Axiom of Choice from the Ordering Principle

In this section, we prove:

THEOREM 5.21. *The Axiom of Choice is independent from the Ordering Principle.*

We will do this by showing that in the basic Cohen model, every set can be linearly ordered.

Let $\mathscr{M}$ be a transitive model of ZF+AC, and let $\mathscr{N}$ be the symmetric extension of $\mathscr{M}$ defined in Section 5.3. We shall keep the notation introduced in Section 5.3; in particular, A is the set of reals

$$A = \{x_n : n \in \omega\},$$

where x_n are defined in (5.11).

We shall make one assumption which simplifies matters for us a little. We shall assume that $\mathscr{M}$ is a class of the model $\mathscr{N}$. This is the case, e.g., when $\mathscr{M} = L$, but in general it need not be true. More generally, let us say that a class $C \subseteq$ HS of names is *symmetric* if sym $(C) \in \mathscr{F}$, where

$$\text{sym}(C) = \{\pi \in \mathscr{G} : \pi''C = C\}.$$

We make the assumption that for each symmetric class C its interpretation $i_G(C) = \{i_G(x) : x \in C\}$ is a class of the model $\mathscr{N}$. As was the case with permutation models, this assumption enables us to formulate statements and construct sets in $\mathscr{N}$ uniformly. E.g., we will be able to show that in the basic Cohen model the universe can be linearly ordered; this is the uniform version of the statement that every set can be linearly ordered.

In Chapter 4, when we proved the analogue of Theorem 5.21 for set theory with atoms, we used the ordered Mostowski model and the crucial property of the model was that each set in the model has a least support. A similar idea will be employed in the proof of Theorem 5.21.

Let $\underset{\sim}{x} \in \mathrm{HS}$ be a symmetric name, and let e be a finite subset of ω. We say that e is a *support* of $\underset{\sim}{x}$ if

$$\mathrm{sym}\,(\underset{\sim}{x}) \supseteq \mathrm{fix}\,(e) = \{\pi \in \mathscr{G} : \pi n = n \text{ for each } n \in e\}.$$

Consider the canonical names $\underset{\sim}{x}_n$ for elements of A, defined in formula (5.12). Each finite $E \subseteq A$ has a canonical name $\underset{\sim}{E}$:

$$\underset{\sim}{E}(\underset{\sim}{x}_n) = 1 \quad \text{for all } x_n \in E.$$

If $\underset{\sim}{x} \in \mathrm{HS}$, let us say that $\underset{\sim}{E}$ is a *support* of $\underset{\sim}{x}$ whenever $E = \{x_{n_1}, \ldots, x_{n_k}\}$ and $\{n_1, \ldots, n_k\}$ supports $\underset{\sim}{x}$. Similarly as in Lemma 4.4, we can see that the class

$$\underset{\sim}{\Delta} = \{(\underset{\sim}{E}, \underset{\sim}{x})^B : \underset{\sim}{E} \text{ is a support of } \underset{\sim}{x}\},$$

is a symmetric class. Thus the relation $\Delta(E, x)$ (where E is a finite subset of A and $x \in \mathscr{N}$), the interpretation of $\underset{\sim}{\Delta}$, is a class of the model $\mathscr{N}$. If $\Delta(E, x)$, we say that E is a *support* of x. The crucial step in the present proof is:

LEMMA 5.22. *Every $x \in \mathscr{N}$ has a least support.*

This lemma follows quite easily from:

LEMMA 5.23. *Let $\underset{\sim}{x}, \underset{\sim}{y}$ be elements of* HS, *let p be a condition and assume that $p \Vdash \underset{\sim}{x} = \underset{\sim}{y}$. If e_1 and e_2 are finite subsets of ω and if e_1 is a support of $\underset{\sim}{x}$ and e_2 is a support of $\underset{\sim}{y}$, then there exists $\underset{\sim}{z} \in$* HS *such that $e_1 \cap e_2$ is a support of $\underset{\sim}{z}$ and $p \Vdash \underset{\sim}{x} = \underset{\sim}{z}$.*

To see that Lemma 5.23 implies Lemma 5.22, let E_1 and E_2 be supports of $x \in \mathscr{N}$. There exists $\underset{\sim}{x}$ such that $\underset{\sim}{E}_1$ is a support of $\underset{\sim}{x}$ and $i_G(\underset{\sim}{x}) = x$, and there exists $\underset{\sim}{y}$ such that $\underset{\sim}{E}_2$ is a support of $\underset{\sim}{y}$ and $i_G(\underset{\sim}{y}) = x$. Hence for some $p \in G$, $p \Vdash \underset{\sim}{x} = \underset{\sim}{y}$, and applying Lemma 5.23, it follows that $E_1 \cap E_2$ is a support of x. Thus it suffices to prove Lemma 5.23.

Before we proceed with the proof, we will prove another lemma. Let p be a forcing condition, i.e., a function defined on a finite subset of $\omega \times \omega$ with values 0 and 1. Let e be a finite subset of ω. We let

$$(5.18) \qquad p : e = \text{the restriction of } p \text{ to } e \times \omega.$$

Clearly, $p : e$ is a weaker condition than p, $p : e \subseteq p$. Let $u \in B = \mathrm{RO}(P)$.

We let

$$u : e = \sum \{(p : e) : p \leqslant u\}. \tag{5.19}$$

Again we have $u : e \geqslant u$. Finally, let

$$B_e = \{(u : e) : u \in B\}. \tag{5.20}$$

It is easily seen that B_e is a complete subalgebra of B.

LEMMA 5.24. *Let $\phi(x_1, ..., x_n)$ be a formula. If $\underline{x}_1, ..., \underline{x}_n \in \mathrm{HS}$ and if e is a support for each of the $\underline{x}_1, ..., \underline{x}_n$, then for each condition p,*

$$\textit{if } p \Vdash \phi(\underline{x}_1, ..., \underline{x}_n) \textit{ then } p : e \Vdash \phi(\underline{x}_1, ..., \underline{x}_n).$$

Hence $[\![\phi(\underline{x}_1, ..., \underline{x}_n)]\!] \in B_e$.

(Note that in particular $[\![\phi(\check{x}_1, ..., \check{x}_n)]\!]$ is either 0 or 1 for any $x_1, ..., x_n \in \mathcal{M}$. This kind of homogeneity is of course a specific property of this particular Cohen model and is not true for every Boolean-valued model.)

PROOF. It suffices to show that no condition q stronger than $p : e$ forces $\neg \phi(\vec{x})$. Let $q \supseteq p : e$; let e_1, e_2 be finite subsets of ω such that $\mathrm{dom}(p) \subseteq e_1 \times \omega$ and $\mathrm{dom}(q) \subseteq e_2 \times \omega$. There exists a permutation $\pi \in \mathcal{G}$ such that $\pi \in \mathrm{fix}(e)$ and $e_1 \cap \pi'' e_2 = e$ (visualize!). It is obvious that p and $\pi(q)$ are compatible. Assuming $q \Vdash \neg \phi(\underline{x})$, we have $\pi q \Vdash \neg \phi(\pi \underline{x})$, and because $\pi \in \mathrm{fix}(e)$, we have $\pi q \Vdash \neg \phi(\underline{x})$, which is a contradiction since $p \Vdash \phi(\underline{x})$ and p and πq are compatible.

PROOF OF LEMMA 5.23. First note that every symmetric *name* has a least support $e \subseteq \omega$. The proof is easier than in Lemma 4.5 because we can use arbitrary permutations of ω. For each $\underline{w} \in \mathrm{HS}$, let

$$s(\underline{w}) = \text{the least support of } \underline{w}. \tag{5.21}$$

Note that for each $\pi \in \mathcal{G}$,

$$s(\pi \underline{w}) = \pi''(s(\underline{w})). \tag{5.22}$$

If p is a forcing condition, let

$$s(p) = \{n \in \omega : (n, m) \in \mathrm{dom}(p) \text{ for some } m\}. \tag{5.23}$$

Before starting the proof, notice one more thing: If $\underline{w} \in \mathrm{HS}$ and $e = s(\underline{w})$, then for any two permutations $\pi, \rho \in \mathcal{G}$,

$$\text{if } \pi | e = \rho | e, \quad \text{then} \quad \pi(\underline{w}) = \rho(\underline{w}). \tag{5.24}$$

(This is because $\rho^{-1}\pi \in \text{fix } e$.) Thus $\pi(\underset{\sim}{w})$ depends only on the restriction of π to $s(\underset{\sim}{w})$.

To prove Lemma 5.23, let $e_1 = s(\underset{\sim}{x})$, $e_2 = s(\underset{\sim}{y})$ and let $p \Vdash \underset{\sim}{x} = \underset{\sim}{y}$. By virtue of Lemma 5.24, we may actually assume that

$$(5.25) \qquad s(p) \subseteq e_1 \cup e_2 .$$

We are looking for a $\underset{\sim}{z} \in \text{HS}$ such that $s(\underset{\sim}{z}) \subseteq e_1 \cap e_2$ and $p \Vdash \underset{\sim}{x} = \underset{\sim}{z}$.

We shall make one assumption which simplifies the arguments somewhat: we assume that $e_1 \cap e_2 = 0$. At the end of the proof, we will discuss the general case $e_1 \cap e_2 \neq 0$ (whose proof is nothing but a 'parametric' version of the present proof). Thus we wish to construct $\underset{\sim}{z} \in \text{HS}$ such that $s(\underset{\sim}{z}) = 0$ and $p \Vdash \underset{\sim}{x} = \underset{\sim}{z}$.

To begin with, we prove that for each $\underset{\sim}{w} \in \text{HS}$,

$$(5.26) \qquad p \cdot [\![\underset{\sim}{w} \in \underset{\sim}{x}]\!] = p \cdot ([\![\underset{\sim}{w} \in \underset{\sim}{x}]\!] : s(\underset{\sim}{w})).$$

To show that (5.26) holds, let $e = s(\underset{\sim}{w})$; it suffices to show that for each $q \leqslant p$,

$$(5.27) \qquad \begin{aligned} &\text{if} \quad q \Vdash \underset{\sim}{w} \in \underset{\sim}{x}, \quad \text{then} \quad p \cup (q : e) \Vdash \underset{\sim}{w} \in \underset{\sim}{x}, \\ &\text{if} \quad q \Vdash \underset{\sim}{w} \notin \underset{\sim}{x}, \quad \text{then} \quad p \cup (q : e) \Vdash \underset{\sim}{w} \notin \underset{\sim}{x}. \end{aligned}$$

Let us prove the first part of (5.27). By a repeated application of Lemma 5.24 and the fact that $p \Vdash \underset{\sim}{x} = \underset{\sim}{y}$, we get for $q \leqslant p$:

$$\begin{aligned} q &\Vdash \underset{\sim}{w} \in \underset{\sim}{x}; \\ q &\Vdash \underset{\sim}{w} \in \underset{\sim}{y}; \\ q : (e \cup e_2) &\Vdash \underset{\sim}{w} \in \underset{\sim}{y}; \\ p \cup (q : e \cup e_2) &\Vdash \underset{\sim}{w} \in \underset{\sim}{x}; \\ (p \cup (q : e \cup e_2)) : (e \cup e_1) &\Vdash \underset{\sim}{w} \in \underset{\sim}{x}; \\ p \cup (q : (e \cup e_2) \cap (e \cup e_1)) &\Vdash \underset{\sim}{w} \in \underset{\sim}{x}; \\ p \cup (q : e) &\Vdash \underset{\sim}{w} \in \underset{\sim}{x}. \end{aligned}$$

Now we shall define $\underset{\sim}{z}$. For each $\underset{\sim}{w} \in \text{HS}$ with $\rho(\underset{\sim}{w}) < \rho(\underset{\sim}{x})$, let

$$(5.28) \qquad \sigma(\underset{\sim}{w}) = \{\pi \underset{\sim}{w} : \pi \in \mathscr{G}\}.$$

In each $\sigma(\underset{\sim}{w})$, pick one $\underset{\sim}{w}$ such that its support $e = s(\underset{\sim}{w})$ satisfies

$$(5.29) \qquad e \cap (e_1 \cup e_2) = 0,$$

and let

$$(5.30)\qquad \underline{z}(\underline{w}) = [\![\underline{w} \in \underline{x}]\!] : e$$

and then

$$(5.31)\qquad \underline{z}(\pi\underline{w}) = \pi(\underline{z}(\underline{w})),$$

for each $\pi \in \mathscr{G}$. The condition (5.31) will guarantee that $\pi\underline{z} = \underline{z}$ for each $\pi \in \mathscr{G}$, as soon as we know that the definition (5.31) is unambiguous; that is, we have to show that $\rho_1(\underline{z}(\underline{w})) = \rho_2(\underline{z}(\underline{w}))$ whenever $\rho_1\underline{w} = \rho_2\underline{w}$. It suffices to show that

$$(5.32)\qquad \text{if}\quad \pi\underline{w} = \underline{w}, \quad\text{then}\quad \pi(\underline{z}(\underline{w})) = \underline{z}(\underline{w})$$

(consider $\pi = \rho_2^{-1}\rho_1$). By (5.30), $\underline{z}(\underline{w})$ belongs to B_e and so the value of $\pi(\underline{z}(\underline{w}))$ depends only on the restriction of π to e. By (5.22), $\pi''e = e$ and since $e \cap e_1 = 0$, we may assume that $\pi \in \operatorname{fix}(e_1)$. Hence $\pi\underline{x} = \underline{x}$ and so

$$\pi[\![\underline{w} \in \underline{x}]\!] = [\![\pi\underline{w} \in \pi\underline{x}]\!] = [\![\underline{w} \in \underline{x}]\!],$$

and by (5.30), we get $\pi(\underline{z}(\underline{w})) = \underline{z}(\underline{w})$.

It remains to show that

$$p \Vdash \underline{x} = \underline{z}.$$

That is, we have to show that for each $\underline{v} \in \mathrm{HS}$ with $\rho(\underline{v}) < \rho(\underline{x})$,

$$(5.33)\qquad p \cdot [\![\underline{v} \in \underline{x}]\!] = p \cdot \underline{z}(\underline{v}).$$

Let $e = s(\underline{v})$. By (5.31), there exists a permutation $\pi \in \mathscr{G}$, and $\underline{w} \in \mathrm{HS}$ such that $\underline{v} = \pi(\underline{w})$, $s(w) = \pi_{-1}(e)$ is disjoint from $e_1 \cup e_2$, and

$$(5.34)\qquad \underline{z}(\underline{v}) = \pi(\underline{z}(\underline{w})) = \pi([\![\underline{w} \in \underline{x}]\!] : s(\underline{w})).$$

Since $\underline{z}(\underline{w}) \in B_{\pi_{-1}(e)}$, the value $\pi(\underline{z}(\underline{w}))$ depends only on the restriction of π to $\pi_{-1}(e)$, and so does $\pi(\underline{w})$ since $s(\underline{w}) = \pi_{-1}(e)$. Hence we may assume that

$$(5.35)\qquad \pi \in \operatorname{fix}(e_1 \cup e_2 - e).$$

On the one hand, we have

$$p \cdot \underline{z}(\underline{v}) = p \cdot ([\![\underline{v} \in \pi\underline{x}]\!]) : e),$$

by (5.34), and on the other hand, by (5.26),

$$p \cdot [\![\underline{v} \in \underline{x}]\!] = p \cdot ([\![\underline{v} \in \underline{x}]\!] : e).$$

Since $[\![\underline{v} \in \underline{x}]\!] : e \in B_e$, it suffices to show: For every $q \leqslant p$ such that $s(q-p) \subseteq e$,

(5.36) (i) if $q \Vdash \underline{v} \in \underline{x}$, then $(\exists r \leqslant q)\ r \Vdash \underline{v} \in \pi\underline{x}$,
(ii) if $q \Vdash \underline{v} \notin \underline{x}$, then $(\exists r \leqslant q)\ r \Vdash \underline{v} \notin \pi\underline{x}$.

Let us prove (5.36)(i). Let $q \leqslant r$ such that $s(q-p) \subseteq e$ and $q \Vdash \underline{v} \in \underline{x}$. We shall show that q and π are compatible and that

$$r = q \cup \pi p \Vdash \underline{v} \in \pi\underline{x}. \tag{5.37}$$

Let $\rho = \pi^{-1}$; we have

$$\rho \in \text{fix}\,(e_1 \cup e_2 - e) \quad \text{and} \quad \rho''e \cap (e_1 \cup e_2) = 0. \tag{5.38}$$

Remembering (5.25) that $s(p) \subseteq e_1 \cup e_2$, it follows that p and ρq are compatible and hence q and πp are compatible. We shall show that

$$\rho p \cup p \Vdash \underline{x} = \rho\underline{x}. \tag{5.39}$$

Then we will have

$$r = \pi(\rho q \cup p) \leqslant \pi(\rho p \cup p) \Vdash \pi\underline{x} = \underline{x}$$
$$r \leqslant q \Vdash \underline{v} \in \underline{x}$$

and (5.37) will follow.

To prove (5.39), let $\sigma \in \mathscr{G}$ be such that

$$\sigma|e_1 = \rho|e_1 \quad \text{and} \quad \sigma \in \text{fix}\,(e_2); \tag{5.40}$$

this is possible because of (5.38). It follows that $\sigma(\underline{x}) = \rho(\underline{x})$, $\sigma(\underline{y}) = \underline{y}$ and $\sigma(p) \subseteq \rho(p) \cup p$. Since $p \Vdash \underline{x} = \underline{y}$, we have

$$\rho p \cup p \Vdash \underline{x} = \underline{y} \wedge \sigma(\underline{x}) = \sigma(\underline{y}),$$

and consequently,

$$\rho p \cup p \Vdash \rho(\underline{x}) = \underline{x}.$$

This completes the proof of Lemma 5.23, under the assumption that $e_1 \cap e_2 = 0$. In general, one has to change several places in the proof to replace 0 by $e_1 \cap e_2$. In particular, (5.26) becomes

$$p \cdot [\![\underline{w} \in \underline{x}]\!] = p \cdot ([\![\underline{w} \in \underline{x}]\!] : (s(\underline{w}) \cup (e_1 \cap e_2))),$$

$\mathscr{G}$ is replaced in (5.28) and (5.31) by fix $(e_1 \cap e_2)$; (5.29) becomes

$$e \cap (e_1 \cup e_2) \subseteq e_1 \cap e_2,$$

etc. We leave this to the reader as an exercise.

Similarly as in Section 4.5, Theorem 5.21 will follow from the following Lemma 5.25 because A is linearly ordered (A is a set of reals) and consequently Lemma 5.25 yields a linear ordering of $\mathcal{N}$ – via the lexicographical ordering of $I \times \mathrm{On}$.

LEMMA 5.25. *In $\mathcal{N}$, there is a one-to-one function $\mathcal{F}$ of $\mathcal{N}$ into $I \times \mathrm{On}$ where I is the set of all finite subsets of A.*

PROOF. The idea is to associate with each $x \in \mathcal{N}$ a finite subset of A (the least support of x) and an element of HS; since $\mathrm{HS} \subseteq \mathcal{M}$ and $\mathcal{M}$ is well-ordered, the lemma will follow. Obviously, one cannot simply choose a name $\underline{x}$ of x, because the interpretation function $i(\underline{x})$ is *not* in $\mathcal{N}$. However, we can define a good approximation of the function i, and this one is in $\mathcal{N}$. (For a general version of Lemma 5.26, see Problem 23.)

We say that t is an *assignment* if t is a one-to-one function whose domain is a finite subset of ω and whose values are elements of A. Let t be an assignment,

$$t(n_1) = x_{i_1}, \ldots, t(n_k) = x_{i_k}, \tag{5.41}$$

and let $\underline{x} \in \mathrm{HS}$ be such that $\{n_1, \ldots, n_k\} \subseteq s(x)$. If π is any permutation such that

$$\pi(n_1) = i_1, \ldots, \pi(n_k) = i_k,$$

then by (5.24), $\pi\underline{x}$ is unique regardless of the behavior of π outside $\{n_1, \ldots, n_k\}$. Let us define

$$e(t, \underline{x}) = i(\pi\underline{x}). \tag{5.42}$$

LEMMA 5.26. *The function e is in $\mathcal{N}$.*

PROOF. There are canonical names for the assignments (using the names $\underline{x}_n$ for $x_n \in A$): $\underline{t}$ is a name for (5.41), and

$$[\![\underline{t}(\check{n}_1) = \underline{x}_{i_1}]\!] = 1, \text{ etc.}. \tag{5.43}$$

The function e is a result of the interpretation by i of a symmetric class $\underline{e}$ consisting of Boolean pairs

$$((\underline{t}, (\underline{x})^{\vee})^B, \pi\underline{x})^B,$$

where $\pi n_1 = i_1$ etc.. The class $\underline{e}$ is symmetric since if $\rho \in \mathcal{G}$, then

$$[\![\rho\underline{t}(n_1) = \underline{x}_{\rho i_1}]\!] = 1, \text{ etc.},$$

$\rho((\underline{x})^{\vee}) = (\underline{x})^{\vee}$ and $\rho(\pi\underline{x}) = (\rho\pi)(\underline{x})$, and $\rho\pi n_1 = \rho i_1$, etc..

Notice that whenever t is an assignment and $\underline{x} \in \mathrm{HS}$ such that $|t| \geqslant s(\underline{x})$, then there is $\underline{y} \in \mathrm{HS}$ such that $i(\underline{x}) = e(t, \underline{y})$; namely, $\underline{y} = \pi\underline{x}$ for some $\pi \in \mathcal{G}$.

Now we define (in $\mathcal{N}$) a one-to-one mapping of $\mathcal{N}$ into $I \times \text{On}$. For $x \in \mathcal{N}$, let $F_1(x) = E$ be the least support of x; let E have k elements $x_{n_1} < x_{n_2} < \ldots < x_{n_k}$ (in that order). Let t be the assignment

$$t(1) = x_{n_1}, \qquad t(2) = x_{n_2}, \ldots, t(k) = x_{n_k}.$$

There exists $\underline{y} \in \text{HS}$ such that

$$x = e(t, \underline{y}).$$

Assuming that we have a fixed enumeration of HS by ordinals, let $F_2(x)$ be the least ordinal corresponding to such $\underline{y}$. Let $F(x) = (F_1(x), F_2(x))$; clearly, if $F(x) = F(y)$, then $x = y$ and so F is one-to-one.

5.6. Problems

If B is a Boolean algebra and $\mathcal{F}$ a family of subsets of B, then a filter G on B is *$\mathcal{F}$-complete* if $\prod \{a : a \in A\} \in G$ whenever $A \subseteq G$, $A \in \mathcal{F}$ and $\prod A$ exists.

1. If B is a complete Boolean algebra in the ground model $\mathcal{M}$, then an ultrafilter G on B is $\mathcal{M}$-generic if and only if it is $\mathcal{P}^{\mathcal{M}}(B)$-complete.

2. Let B be as in Problem 1, let $\mathcal{F} = \mathcal{P}^{\mathcal{M}}(B)$ and let $\underline{G}$ be the canonical generic ultrafilter in $\mathcal{M}^B$. Then

$$[\![\underline{G} \text{ is an } \check{\mathcal{F}}\text{-complete ultrafilter on } \check{B}]\!] = 1.$$

[*Hint*: $[\![A^{\vee} \subseteq G]\!] = [\![(\prod A)^{\vee} \in G]\!] = \prod A$.]

A partial ordering is *separative* if for all p, q such that $p \not\leqslant q$, there exists $r \leqslant p$ which is incompatible with q.

3. For every partially ordered set $(P, <)$ there exists a unique (up to isomorphism) separative partially ordered set $(Q, \prec)$ and a homomorphism h of P onto Q such that for all $p, q \in P$,

$$p, q \text{ are compatible in } P \leftrightarrow hp, hq \text{ are compatible in } Q.$$

[*Hint*: Let $p \preccurlyeq q$ if q is compatible with each $x \leqslant p$.]

4. Every separative partial ordering $(Q, \leqslant)$ can be embedded isomorphically onto a dense subset of a (unique) complete Boolean algebra.

[*Hint*: Consider subsets A of Q with the following properties: (a) if $y \leqslant x \in A$, then $y \in A$, (b) if $(\forall y \leqslant x)(\exists z \leqslant y)\,[z \in A]$, then $x \in A$. The family of all such sets, partially ordered by inclusion, is a complete Boolean algebra.]

5. Let κ be a cardinal and assume that for every descending sequence $p_0 \geqslant p_1 \geqslant \ldots \geqslant p_\xi \geqslant \ldots$ $(\xi < \kappa)$ of conditions, there exists $p \in P$ such that $p \leqslant p_\xi$ for all $\xi < \kappa$. Then if $f \in \mathcal{M}[G]$ is a function from κ into $\mathcal{M}$, then $f \in \mathcal{M}$. Consequently, $\mathcal{M}[G]$ has the same cardinals below κ as $\mathcal{M}$ (including κ).

[*Hint*: As in Lemma 5.12. The intersection of κ open dense sets is open dense.]

A partially ordered set $(P, <)$ satisfies *the countable chain condition* if every set of pairwise incompatible elements of P is countable. P satisfies the κ-*chain condition* (where κ is a cardinal) if every set of pairwise incompatible elements of P has cardinality less than κ.

6. $(P, <)$ has the κ-chain condition if and only if $\mathrm{RO}(P)$ has the κ-chain condition.

*7. Let κ be a regular cardinal and assume that $B \in \mathcal{M}$ satisfies the κ-chain condition. Then κ is a regular cardinal in $\mathcal{M}[G]$.

[*Hint*: If $[\![\check{\kappa}$ is not a regular cardinal$]\!] \neq 0$, then there exists a family $\{w(\alpha, \beta) : \alpha < \lambda, \beta < \kappa\}$, where $\lambda < \kappa$, of elements of B such that (a) $w(\alpha, \beta) \cdot w(\alpha, \beta_1) = 0$ if $\beta \neq \beta_1$ and (b) $\forall \beta\, (\exists \beta_1 > \beta)\, \exists \alpha\, [w(\alpha, \beta_1) \neq 0]$. This contradicts the κ-chain condition.]

Consequently, if B satisfies the countable chain condition in $\mathcal{M}$, then every cardinal in $\mathcal{M}$ is a cardinal in $\mathcal{M}[G]$.

*8. Assume that $\mathcal{M}$ satisfies the Generalized Continuum Hypothesis. Let $\aleph_\alpha$ be a regular cardinal in $\mathcal{M}$. Let $(P, <)$ be the set of all transfinite 0-1-sequences of length less than ω_α; the partial ordering is by the inverse inclusion. Then every cardinal in $\mathcal{M}$ is a cardinal in $\mathcal{M}[G]$, and $\aleph_\alpha$ is the least cardinal which has a new subset in $\mathcal{M}[G]$.

[*Hint*: For cardinals below $\aleph_\alpha$, use Problem 5. For cardinals above $\aleph_\alpha$, use Problem 7: $(P, <)$ satisfies the $\aleph_{\alpha+1}$-chain condition.]

9. Let $(P, <)$ be the set of all functions p whose domain is a set $\{0, 1, \ldots, n-1\}$ for some n, with values being countable ordinals; the partial ordering is by the inverse inclusion (in $\mathcal{M}$). Then $(\omega_1)^{\mathcal{M}}$ is a countable ordinal in $\mathcal{M}[G]$.

[*Hint*: A generic set of conditions yields a function from ω onto ω_1.]

*10. Let $(P, <)$ be the set of all functions p whose domain is a finite subset of $\omega \times \omega_2$ with values 0 and 1; the partial ordering is by the inverse inclusion. In $\mathcal{M}[G]$, $2^{\aleph_0} > \aleph_1$.

[*Hint*: $\bigcup \{p : p \in G\}$ is a function from $\omega \times \omega_2$ into $\{0, 1\}$. For $\alpha < \omega_2$, let $z_\alpha = \{n \in \omega : F(n, \alpha) = 1\}$; show that $[\![\underline{z}_\alpha = \underline{z}_\beta]\!] = 0$ for $a \neq \beta$. Show that $(P, <)$ satisfies the countable chain condition; thus $\mathcal{M}[G]$ has the same cardinals as $\mathcal{M}$.]

11. Let G be $\mathcal{M}$-generic on $B \in \mathcal{M}$; let κ be a cardinal in $\mathcal{M}[G]$. Then $(2^\kappa)^{\mathcal{M}[G]} \leqslant (|B|^\kappa)^{\mathcal{M}}$.

[*Hint*: If $A \subseteq \kappa$ in $\mathcal{M}[G]$ and if $\underline{A}$ is a name of A, then $\underline{A}$ represents a function $\alpha \mapsto [\![\check{\alpha} \in \underline{A}]\!]$ from κ into B.]

12. If $\mathcal{M}$ satisfies the Generalized Continuum Hypothesis, then in $\mathcal{M}[G]$ there is some α_0 such that $2^{\aleph_\alpha} = \aleph_{\alpha+1}$ for all $\alpha \geqslant \alpha_0$.

[*Hint*: Use Problem 11.]

**13. Assume that $\mathcal{M}$ satisfies the Generalized Continuum Hypothesis. Let F be a function (in $\mathcal{M}$) from regular cardinals into cardinals such that (a) if $\alpha \leqslant \beta$ then $F(\alpha) \leqslant F(\beta)$ and (b) $\operatorname{cf}(F(\alpha)) > \alpha$. There exists a model $\mathcal{N}$ of ZF+AC such that $\mathcal{N} \supseteq \mathcal{M}$, $\mathcal{N}$ has the same cardinals as $\mathcal{M}$, and $(2^\alpha)^{\mathcal{N}} = F(\alpha)$.

The construction of the model $\mathcal{N}$ is a generalization of generic models defined in Section 5.1. (Note that by Problem 12, $\mathcal{N}$ cannot be a generic model in the sense of Section 5.1.) Instead of a set of forcing conditions, one has to use a proper class of forcing conditions.

**14. Let $\mathcal{M}$ be a transitive model of ZF+AC. There exists a model $\mathcal{N}$ of ZF such that $\mathcal{N} \supseteq \mathcal{M}$, $\mathcal{N}$ has the same cardinals as $\mathcal{M}$, and the following statement is true in $\mathcal{N}$: For each α there exists a set X such that X is a countable union of countable sets and $\mathcal{P}(X)$ can be partitioned into $\aleph_\alpha$ nonempty sets.

Notice that there is no transitive model $\mathcal{M}_1$ of ZF+AC such that $\mathcal{N}$ is a submodel of $\mathcal{M}_1$ with the same ordinals: $\mathcal{M}_1$ would satisfy that $2^{\aleph_0}$ can be partitioned into any number of nonempty sets.

Let $\mathcal{N}$ be a symmetric extension of $\mathcal{M}$ given by B, $\mathcal{G}$ and $\mathcal{F}$. If H is a subgroup of $\mathcal{G}$, let $\mathrm{b}H = \{u \in B : \pi u = u \text{ for each } \pi \in H\}$; $\mathrm{b}H$ is a complete subalgebra of B. Let $\mathcal{I}(\mathcal{F})$ be the ideal of complete subalgebras of B generated by $\{\mathrm{b}H : H \in \mathcal{F}\}$.

*15. If $\mathcal{F}_1$ and $\mathcal{F}_2$ are two normal filters on $\mathcal{G}$, if $\mathcal{N}_1$ and $\mathcal{N}_2$ are the two symmetric extensions of $\mathcal{M}$ given by $\mathcal{F}_1$ and $\mathcal{F}_2$ respectively (B and G are the same), and if $\mathcal{I}(\mathcal{F}_1) = \mathcal{I}(\mathcal{F}_2)$, then $\mathcal{N}_1$ and $\mathcal{N}_2$ have the same sets of ordinal numbers.

[*Hint*: Each set of ordinals $A \in \mathcal{M}[G]$ has a name $\underline{A}$ whose domain consists of $\check{\alpha}$ for some ordinals α. Let $B(\underline{A})$ be the complete subalgebra of B generated by $\{\underline{A}(\alpha) : \check{\alpha} \in \operatorname{dom} \underline{A}\}$. Note that $\operatorname{sym}(\underline{A}) = \operatorname{fix}(B(\underline{A})) = \{\pi \in \mathcal{G} : \pi u = u \text{ for each } u \in B(\underline{A})\}$.]

Compare the following result with Problem 3.19.

*16. There exist two different symmetric extension $\mathcal{N}_1, \mathcal{N}_2$ of $\mathcal{M}$ which have the same sets of ordinal numbers.

[*Hint*: $(P, <)$ consists of finite functions p with values 0, 1 and $\operatorname{dom}(p) \subseteq \omega \times \omega$. Let $\underline{x}_n$ and $\underline{A}$ be as defined in (5.12) and (5.13). Every permutation π of ω induces an automorphism of RO(P) via $(\pi p)(\pi n, m) = p(n, m)$ as in Section 5.3. Let $\mathcal{G}_1$ be the group of all such permutations; let $\mathcal{F}_1$ be the filter on $\mathcal{G}_1$ generated by the groups $\operatorname{fix}(e)$, $e \subseteq \omega$ finite. (Thus $\mathcal{N}_1$ is the basic Cohen model.)

Every permutation π of $\omega \times \omega$ induces an automorphism of RO(P) via $(\pi p)(\pi(n, m)) = p(n, m)$. Let $\mathcal{G}_2$ be the group of all permutations π of $\omega \times \omega$ which move only finitely many elements; let $\mathcal{F}_2$ be the filter on $\mathcal{G}_2$ generated by $\{\operatorname{fix}(e) : e \subseteq \omega \text{ finite}\}$. One can prove that (a) $\mathcal{F}_2$ is a normal filter, (b) $\mathcal{I}(\mathcal{F}_1) = \mathcal{I}(\mathcal{F}_2)$, and (c) the set A is not in $\mathcal{N}_2$. The last statement is proved by a method analogous to (5.15), for example.]

**17. There exists an ω-sequence of models of ZF such that the nth and $(n+1)^{\text{th}}$ models have the same sets of sets ... (n times) of ordinals.

As a consequence, none of the statements K(n) is provable in ZF alone (see Problem 3.20).

18. The basic Cohen model has an infinite set of real numbers which is Dedekind finite.

[*Hint*: $A = \{x_n : n \in \omega\}$ is the set. The proof is implicit in Lemma 5.15.]

19. If X is an infinite set of reals then $\mathcal{P}(X)$ is Dedekind infinite.

[*Hint*: Let $X \subseteq {}^{\omega}2$ be infinite. Let S be the set of all finite sequences of 0's and 1's. For $s \in S$, let $X_s = \{f \in X : s \text{ is an initial segment of } f\}$. The set $\{X_s : s \in S\}$ is an infinite countable subset of $\mathcal{P}(X)$.]

20. In the basic Cohen model, if X is infinite, then $\mathcal{P}(X)$ is Dedekind infinite.

[*Hint*: By Lemma 5.25, there exists α such that $|X| \leqslant |\mathcal{P}(\omega) \times \alpha|$. Use Problem 19.]

Compare this result with Problem 4.5.

21. The Selection Principle holds in the basic Cohen model.

[*Hint*: Use Lemma 5.25 and Problem 4.12.]

Thus the Axiom of Choice is independent from the Selection Principle. Compare this with Problem 4.13.

22. In the basic Cohen model, every family of nonempty well-orderable sets has a choice function.

[*Hint*: Let $S \in \mathcal{N}$ be well-orderable; there is a one-to-one function f from ordinals to S. Let $X \in S$, let $p \Vdash \underset{\sim}{X} = \underset{\sim}{f}(\check{\alpha})$. Let e be a support of $\underset{\sim}{f}$, e_1 a support of $\underset{\sim}{X}$; we may assume $s(p) \subseteq e \cup e_1$. Let $\pi \in \text{fix}\,(e)$ such that $e \cap \pi e_1 = e \cap e_1 = e_1 \cap \pi e_1$. For each $q \leqslant p$ such that $s(q) \subseteq e \cup e_1$, $q \cup \pi q \Vdash \underset{\sim}{X} = \underset{\sim}{f}(\check{\alpha}) = \pi \underset{\sim}{X}$; thus $p \Vdash \underset{\sim}{X} = \pi \underset{\sim}{X}$. By Lemma 5.23, the least support of X is included in the least support of f. Hence for some finite E, we have $S \subset \Delta(E)$, the class of all x such that E is a support of x. The rest is as in Problem 4.14.]

The following is a general version of Lemma 5.26.

23. Let I be a subset of HS such that:

(a) $\{\text{sym}\,(\underset{\sim}{E}) : \underset{\sim}{E} \in I\}$ generates $\mathscr{F}$;

(b) if $\underset{\sim}{E} \in I$ and $\pi \in \mathscr{G}$, then $\pi(\underset{\sim}{E}) \in I$;

(c) if $\underset{\sim}{E}_1 \neq \underset{\sim}{E}_2 \in I$, then $[\![\underset{\sim}{E}_1 = \underset{\sim}{E}_2]\!] = 0$.

An *assignment* is a pair $t = (\underset{\sim}{E}, i(\pi \underset{\sim}{E}))$ where $\underset{\sim}{E} \in I$ and $\pi \in \mathscr{G}$. If $\underset{\sim}{x} \in \text{HS}$ and $t = (\underset{\sim}{E}, i(\pi \underset{\sim}{E}))$, and if $\text{sym}\,(\underset{\sim}{E}) \subseteq \text{sym}\,(\underset{\sim}{x})$, then we define

$$e(t, \underset{\sim}{x}) = i(\pi \underset{\sim}{x}).$$

THEOREM: *The function e is in* $\mathcal{N}$.

24. Let $(P, <)$ be the set of all finite functions p with values 0, 1, and $\text{dom}\,(p) \subseteq \omega \times \omega$. Every $X \subseteq \omega \times \omega$ induces an automorphism π_X of $\text{RO}(P)$ via $(\pi p)(n, m) =$ either $p(n, m)$ or $1 - p(n, m)$, according to whether $(n, m) \notin X$ or $(n, m) \in X$. Let $\mathscr{G} = \{\pi_X : X \subseteq \omega \times \omega\}$. Let $\mathscr{F}$ be the filter on $\mathscr{G}$ generated by $\{\text{fix}\,(e) : e \subseteq \omega \text{ finite}\}$, where $\text{fix}\,(e) = \{\pi_X : X \cap (e \times \omega) = 0\}$. Let $\mathcal{N}$ be the corresponding symmetric model.

THEOREM. *In* $\mathcal{N}$, *there is no nontrivial ultrafilter over* ω.

[*Hint*: Let $\underset{\sim}{D} \in \text{HS}$ and let $p \Vdash \underset{\sim}{D}$ be an ultrafilter over $\check{\omega}$; let $\text{sym}\,(\underset{\sim}{D}) \supseteq \text{fix}\,(e)$ for some finite $e \subseteq \omega$. Let $n \notin e$ and let x_n be the subset of ω defined in (5.11). If $q \Vdash \underset{\sim}{x}_n \in \underset{\sim}{D}$, let m_0 be such that for each $m \geqslant m_0, (n, m) \notin \text{dom}\,(q)$; let $X = \{(nm), m \geqslant m_0\}$. Then we have $\pi_X q = q \Vdash \pi_X(x_n) \in \underset{\sim}{D}$. Similarly for $q \Vdash \underset{\sim}{x}_n \notin D$. Since $x_n \cap i(\pi \underset{\sim}{x}_n) \subseteq m_0$, D must be trivial.]

*25. *Independence of the Hahn–Banach Theorem.* Modify the construction of the model $\mathcal{N}$ in Problem 24 as follows: in addition to the 'symmetries' π_X, consider those 'permutations of $\omega\times\omega$' which move only elements in finitely many columns (compare with Problem 16). In the resulting model, there is no measure on the Boolean algebra $\mathscr{P}(\omega)/I$, where I is the ideal of finite sets. Therefore (cf. Problem 2.19), the Hahn–Banach Theorem fails in the model.

[*Hint*: Let $\underline{\mu} \in \mathrm{HS}$ be a measure on $\mathscr{P}(\omega)$, vanishing on finite sets, with $\mathrm{sym}(\underline{\mu}) \supseteq \mathrm{fix}(e)$. The same argument as in Problem 24 gives $\mu(x_n) = \mu(\omega - x_n)$ whenever $n \notin e$; therefore $\mu(x_n) = \frac{1}{2}$. Then use a permutation π which maps $\underline{x}_{n_1}$ and $\underline{x}_{n_2}$ (up to a finite set) onto disjoint subsets of $\underline{x}_{n_3}$ and such that $\pi\underline{\mu} = \underline{\mu}$. A contradiction.]

It should be mentioned that the models in Problems 24 and 25, as well as the model $\mathcal{N}_2$ in Problem 16 are the same. This model is the least submodel of the basic Cohen model which contains all the sets x_n (but not the collection $A = \{x_n : n \in \omega\}$ of these sets).

*26. Let $(P, <)$ be the set of all finite functions p with values 0, 1, and $\mathrm{dom}(p) \subseteq \omega\times\omega_1$. Let $\mathcal{M}[G]$ be the corresponding generic extension. In $\mathcal{M}[G]$, let $\mathcal{N}$ be the class of all sets which are hereditarily definable from a countable sequence of ordinals. (The actual construction is as for HOD sets, cf. Problem 3.21.) $\mathcal{N}$ is a model of ZF (see Problem 3.23). In $\mathcal{N}$, the Axiom of Choice fails, and the Principle of Dependent Choices is true. Thus the Axiom of Choice is independent of the Principle of Dependent Choices.

[*Hint*: (a). The Axiom of Choice fails in $\mathcal{N}$. The set of all reals cannot be enumerated by ordinals. Using the Countable Chain Condition, show that every countable sequence of ordinals in $\mathcal{M}[G]$ has a name which involves only countably many conditions. Thus every $x \in \mathcal{N}$ has a name which involves only countably many conditions (use Lemma 5.24, which holds for this model also). Similarly as in (5.11) and (5.12), one can define reals x_α, $\alpha < \omega_1$. If $f \in \mathcal{N}$ is a function from the reals into ordinals, then we can find $\alpha \neq \beta$ such that $f(x_\alpha) = f(x_\beta)$; here we use permutations and Lemma 5.13.

(b) The Principle of Dependent Choices holds in $\mathcal{N}$. First show that if $f \in \mathcal{M}[G]$ is a function from ω into $\mathcal{N}$, then $f \in \mathcal{N}$. The proof goes roughly as follows: For each n, $f(n)$ is definable from some countable sequence of ordinals, choose one such s_n. The sequence of sequences $\langle s_0, s_1, s_2, \ldots\rangle$ can be coded by a single countable sequence of ordinals. Thus f is definable from a countable sequence of ordinals and so $f \in \mathcal{N}$. If $\rho \in \mathcal{N}$ is a relation on $A \in \mathcal{N}$ such that $(\forall x \in A)(\exists y \in A)[x \rho y]$, then there is a sequence $\langle x_0, x_1, x_2, \ldots \rangle = f$ in $\mathcal{M}[G]$ such that $x_0 \rho x_1$, $x_1 \rho x_2$ etc. By the preceding claim, f is in $\mathcal{N}$.]

5.7. Historical remarks

The method of forcing was invented by Cohen [1963a, b, 1964, 1966], who also proved the independence of the Continuum Hypothesis and of the Axiom of Choice, and constructed the models in Sections 5.1, 5.3 and 5.4, and in Problems 8, 9 and 10. The Boolean-valued version is due to Scott and Solovay [1967] and Vopěnka [1967]. The proof of the independence of the Axiom of Choice from the Ordering Principle is due to Halpern and Levy [1964].

The theorem in Problem 13 is due to Easton [1964]. The model in Problem 14 was constructed by Morris [1970]. The result in Problem 16 is due to Jech [1967]; Problem 17 is due to Monro [1972]. The observation in Problem 20 is from Pincus' thesis [1969]. Problem 21 is a part of the abovementioned result of Halpern and Levy. The statement of Problem 22 was probably first noticed by Mathias [1967]. The theorem in Problem 23 was proved by Feferman [1965]. Problem 25 is due to Pincus [1973b] (also, Solovay [1970] notes that the Hahn–Banach Theorem fails in his model). The construction in Problem 26 is due to Solovay. The independence of the Axiom of Choice from the Principle of Dependent Choices was first announced by Feferman in [1964] (for ZFA see Mostowski [1948]).

CHAPTER 6

EMBEDDING THEOREMS

6.1. The First Embedding Theorem

An observant reader could hardly fail to notice that there is a certain similarity between permutation models of ZFA and symmetric models of ZF. For instance, in the basic Cohen model in Section 5.3, the set A behaves similarly to the set of atoms in a permutation model (either in the basic Fraenkel model or in the linearly ordered Mostowski model). The analogy is, however, far from complete. The set A, being a set of reals, carries already a certain structure and so it must necessarily be different from a set of atoms which in general are unrecognizable one from another. [Notice, e.g., that the Selection Principle holds in the basic Cohen model while it fails in the Mostowski model. Also, $\mathscr{P}(A)$ is Dedekind finite in the Mostowski model, while each $\mathscr{P}(X)$ is either finite or Dedekind infinite in the basic Cohen model; cf. Problems 4.5, 4.13, 5.20 and 5.21.]

The second Cohen model in Section 5.4 suggests that sets of reals, rather than reals, are more suitable to play the role of atoms and this can be utilized to the extent that every permutation model can be simulated by a symmetric model, with atoms being replaced by *sets of sets of ordinals*. The present chapter deals with a general method of transfer and some of its refinements.

THEOREM 6.1 (First Embedding Theorem). *Let $\mathscr{U}$ be a model of* ZFA+AC, *let A be the set of all atoms of $\mathscr{U}$, let $\mathscr{M}$ be the kernel of $\mathscr{U}$ and let α be an ordinal in $\mathscr{U}$. For every permutation model $\mathscr{V} \subseteq \mathscr{U}$ (a model of* ZFA), *there exists a symmetric extension $\mathscr{N} \supseteq \mathscr{M}$ (a model of* ZF) *and a set $\tilde{A} \in \mathscr{N}$ such that*

$$(\mathscr{P}^{\alpha}(A))^{\mathscr{V}} \text{ is } \in\text{-isomorphic to } (\mathscr{P}^{\alpha}(\tilde{A}))^{\mathscr{N}}.$$

PROOF. Let α be an ordinal in $\mathscr{U}$, let $\mathscr{G}, \mathscr{F} \in \mathscr{U}$ be a group of permutations of the set A of all atoms of $\mathscr{U}$ and a normal filter on $\mathscr{G}$, respectively. Let $\mathscr{V}$ be the class of all hereditarily symmetric elements of $\mathscr{U}$ (w.r.t. $\mathscr{G}, \mathscr{F}$) and let $\mathscr{M}$ be the kernel of $\mathscr{U}$. First we construct a generic extension $\mathscr{M}[G]$ of $\mathscr{M}$.

Let κ be a regular cardinal such that $\kappa > |\mathscr{P}^{\alpha}(A)|$ (in $\mathscr{U}$). The set P of forcing conditions consists of functions p with values 0, 1 such that $|\text{dom}(p)| < \kappa$ and $\text{dom}(p) \subseteq (A \times \kappa) \times \kappa$; as usual, $\leqslant\ =\ \supseteq$. [In fact, we rather let $\text{dom}(p) \subseteq (A' \times \kappa) \times \kappa$, where $A' \in \mathscr{M}$ has the same cardinality as A, in order to satisfy the formality that $(P, <) \in \mathscr{M}$.]

Let G be an $\mathscr{M}$-generic ultrafilter on $B = \text{RO}(P)$. We define the following elements of $\mathscr{M}[G]$, together with their canonical names in $\mathscr{M}^B$:

$$x_{a\xi} \subseteq \kappa,\ \eta \in x_{a\xi} \leftrightarrow (\exists p \in G)[p(a, \xi, \eta) = 1] \quad (a \in A, \xi < \kappa), \tag{6.1}$$
$$\text{dom}(\underset{\sim}{x}_{a\xi}) = \{\check{\eta} : \eta < \kappa\}, \quad \underset{\sim}{x}_{a\xi}(\check{\eta}) = u_{a\xi\eta} = \sum \{p \in P : p(a, \xi, \eta) = 1\},$$

$$\tilde{a} = \{x_{a\xi} : \xi < \kappa\} \quad (a \in A), \tag{6.2}$$
$$\underset{\sim}{\tilde{a}}(\underset{\sim}{x}_{a\xi}) = 1 \quad \text{for all } \xi < \kappa,$$

$$\tilde{A} = \{\tilde{a} : a \in A\}, \tag{6.3}$$
$$\underset{\sim}{\tilde{A}}(\underset{\sim}{\tilde{a}}) = 1 \quad \text{for all } a \in A.$$

For every $x \in \mathscr{U}$, we define $\tilde{x} \in \mathscr{M}[G]$ and its canonical name $\underset{\sim}{\tilde{x}} \in \mathscr{M}^B$ by $\in$-recursion: if x is a set in $\mathscr{U}$ and if $\tilde{y}$ and $\underset{\sim}{\tilde{y}}$ have been defined for every $y \in x$, let

$$\tilde{x} = \{\tilde{y} : y \in x\},$$
$$\text{dom}(\underset{\sim}{\tilde{x}}) = \{\underset{\sim}{\tilde{y}} : y \in x\}, \quad \underset{\sim}{\tilde{x}}(\underset{\sim}{\tilde{y}}) = 1 \quad \text{for all } y \in x. \tag{6.4}$$

Since $\tilde{a}$ and $\underset{\sim}{\tilde{a}}$ have been defined for all $a \in A$, $\tilde{x}$ and $\underset{\sim}{\tilde{x}}$ are defined for all $x \in \mathscr{U}$.

LEMMA 6.2. *For all $x, y \in \mathscr{U}$,*

$$x \in y \leftrightarrow \tilde{x} \in \tilde{y}, \qquad x = y \leftrightarrow \tilde{x} = \tilde{y}.$$

PROOF. Clearly, if $x \in y$, then $[\![\underset{\sim}{\tilde{x}} \in \underset{\sim}{\tilde{y}}]\!] = 1$ and so $\tilde{x} = \tilde{y}$. To show that $\tilde{x} \in \tilde{y}$ implies $x \in y$ and $\tilde{x} = \tilde{y}$ implies $x = y$, we have to use the induction on ranks, simultaneously for $\in$ and $=$ (simultaneously as in Lemma 5.6, for example). First note that $[\![\underset{\sim}{x}_{a\xi} = \underset{\sim}{x}_{a'\xi'}]\!] = 0$ whenever $(a, \xi) \neq (a', \xi')$, and $[\![\underset{\sim}{x}_{a\xi} = \check{x}]\!] = 0$ for any $x \in \mathscr{M}$; the proof is as in the proof of Lemma 5.15. Consequently, $\tilde{a}_1 \neq \tilde{a}_2$ whenever $a_1 \neq a_2 \in A$. We claim that for all $x \in \mathscr{U}$, $\tilde{x}$ is not equal to any $x_{a\xi}$. If $x \in \mathscr{M}$, then $\tilde{x} = x$ and so $\tilde{x} \neq x_{a\xi}$. If $x \in \mathscr{V} - \mathscr{M}$, then $\tilde{x}$ is of higher rank than any $x_{a\xi}$; viz., $x_{a\xi}$ is a set of ordinals, whereas the transitive closure of $\tilde{x}$ contains some of the $x_{a\xi}$'s. Now we can prove the lemma:

(a). If $\tilde{x} \in \tilde{y}$, then y cannot be an atom, because then we would have $\tilde{x} = x_{a\xi}$ for some a, ξ, which is impossible. Hence $\tilde{x} = \tilde{z}$ for some $z \in y$, and by the induction hypothesis, $x = z$; thus $x \in y$.

(b) If $x \neq y$, then either both x and y are atoms and then $\tilde{x} \neq \tilde{y}$, or, e.g., x contains some z which is not in y, and then by the induction hypothesis, $\tilde{z} \in \tilde{x}$ and $\tilde{z} \notin \tilde{y}$; thus $\tilde{x} \neq \tilde{y}$.

Now we construct a symmetric submodel $\mathcal{N}$ of $\mathcal{M}[G]$; we shall define a group $\overline{\mathcal{G}}$ of automorphisms of B and a normal filter $\overline{\mathcal{F}}$ on $\overline{\mathcal{G}}$.

For every permutation ρ of A, let $[\rho]$ be the group of all permutations π of $A \times \kappa$ such that for all a, ξ,

$$\pi(a, \xi) = (\rho(a), \xi') \quad \text{for some } \xi'.\text{[1]} \tag{6.5}$$

We let

$$\overline{\mathcal{G}} = \bigcup \{[\rho] : \rho \in \mathcal{G}\}; \tag{6.6}$$

similarly, we let $\overline{H} = \bigcup \{[\rho] : \rho \in H\}$ for every subgroup H of $\mathcal{G}$. Since every permutation π of $A \times \kappa$ induces an automorphism π of B via

$$(\pi p)(\pi(a, \xi), \eta) = p(a, \xi, \eta) \quad \text{for all } a, \xi, \eta (p \in P) \tag{6.7}$$
$$\pi u = \sum \{\pi p : p \in P \text{ and } p \leqslant u\} \quad (u \in B),$$

we consider $\overline{\mathcal{G}}$ as a group of automorphisms of B. For every finite $e \subseteq A \times \kappa$, let

$$\text{fix}\,(e) = \{\pi \in \overline{\mathcal{G}} : \pi x = x \text{ for each } x \in e\}. \tag{6.8}$$

We let $\overline{\mathcal{F}}$ be the filter on $\overline{\mathcal{G}}$ generated by

$$\{\overline{H} : H \in \mathcal{F}\} \cup \{\text{fix}\,(e) : e \subseteq A \times \kappa,\ e \text{ finite}\}. \tag{6.9}$$

The reader will gladly verify that $\overline{\mathcal{F}}$ is a normal filter.

Let HS $\subseteq \mathcal{M}^B$ be the class of all hereditarily symmetric names (w.r.t. $\overline{\mathcal{G}}$, $\overline{\mathcal{F}}$), and let the symmetric model $\mathcal{N}$ be the interpretation of HS by G. It is clear that

(a) $x_{a\xi} \in \mathcal{N}$ for all $a \in A$, $\xi < \kappa$, because $\text{sym}_{\overline{\mathcal{G}}}\,(\underline{x}_{a\xi}) = \text{fix}\,\{(a, \xi)\}$;

(b) $\tilde{a} \in \mathcal{N}$ for all $a \in A$, because $\text{sym}_{\overline{\mathcal{G}}}\,(\tilde{a}) = \overline{\text{sym}_{\mathcal{G}}\,(a)}$;

(c) $\tilde{A} \in \mathcal{N}$, because $\text{sym}\,(\underline{\tilde{A}}) = \overline{\mathcal{G}}$.

[1] We recommend that the reader visualizes $A \times \kappa$ as a set A of disjoint blocks, each block consisting of κ elements; ρ moves the blocks while π moves the elements; π acts on the blocks exactly as ρ does.

LEMMA 6.3. *For all* $x \in \mathscr{U}$,

$$x \in \mathscr{V} \leftrightarrow \underset{\sim}{\tilde{x}} \in \mathrm{HS}.$$

PROOF. It suffices to show that

$$\text{(6.10)} \qquad \operatorname{sym}_{\mathscr{G}}(x) \in \mathscr{F} \leftrightarrow \operatorname{sym}_{\bar{\mathscr{G}}}(\underset{\sim}{\tilde{x}}) \in \overline{\mathscr{F}}.$$

If $\rho \in \mathscr{G}$ and $\pi \in [\rho]$, then $\underset{\sim}{\widetilde{\rho x}} = \pi(\underset{\sim}{\tilde{x}})$, and so $\operatorname{sym}_{\bar{\mathscr{G}}}(\underset{\sim}{\tilde{x}}) = \overline{\operatorname{sym}_{\mathscr{G}}(x)}$; thus if $\operatorname{sym}_{\mathscr{G}}(x) \in \mathscr{F}$, then $\operatorname{sym}_{\bar{\mathscr{G}}}(\underset{\sim}{\tilde{x}}) \in \overline{\mathscr{F}}$. On the other hand, if $\operatorname{sym}_{\bar{\mathscr{G}}}(\underset{\sim}{\tilde{x}}) \in \overline{\mathscr{F}}$, then $\overline{\operatorname{sym}_{\mathscr{G}}(x)} \supseteq \bar{H} \cap \operatorname{fix}(e)$ for some $H \in \mathscr{F}$ and finite $e \subseteq A \times \kappa$. If $E \subseteq A$ is the projection of e into A (i.e., $E = \{a \in A : (a, \xi) \in e \text{ for some } \xi\}$), then $\operatorname{sym}_{\mathscr{G}}(x) \supseteq H \cap \operatorname{fix}(E)$ and hence $\operatorname{sym}_{\mathscr{G}}(x) \in \mathscr{F}$.

LEMMA 6.4. *For all* $x \in \mathscr{U}$,

$$x \in \mathscr{V} \leftrightarrow \tilde{x} \in \mathscr{N}.$$

PROOF. In view of Lemma 6.3, it suffices to show that if $\tilde{x} \in \mathscr{N}$, then $x \in \mathscr{V}$. Assume that $x \in \mathscr{U}$ is of least rank such that $\tilde{x} \in \mathscr{N}$ and $x \notin \mathscr{V}$. Thus $x \subseteq \mathscr{V}$; since $\tilde{x} \in \mathscr{N}$, there exists $\underset{\sim}{z} \in \mathrm{HS}$ and $p \in G$ such that $p \Vdash z = \underset{\sim}{\tilde{x}}$. We have $\operatorname{sym}_{\bar{\mathscr{G}}}(z) \in \overline{\mathscr{F}}$ and so there exists $H \in \mathscr{F}$ and a finite $e \subseteq A \times \kappa$ such that $\operatorname{sym}(\underset{\sim}{z}) \supseteq \bar{H} \cap \operatorname{fix}(e)$. We shall find $\rho \in \mathscr{G}$ and $\pi \in [\rho]$ such that:

(6.11) (i) πp and p are compatible;
(ii) $\pi \in \bar{H} \cap \operatorname{fix}(e)$,
(iii) $\rho(x) \neq x$.

Then we will have $\pi \underset{\sim}{z} = \underset{\sim}{z}$ by (6.11)(ii), $[\![\pi(\underset{\sim}{\tilde{x}}) = \underset{\sim}{\tilde{x}}]\!] = 0$ by (6.11)(iii), and since

$$\pi p \Vdash \pi \underset{\sim}{z} = \pi \underset{\sim}{\tilde{x}},$$

we will have

$$q = p \cup \pi p \Vdash (\underset{\sim}{z} = \underset{\sim}{\tilde{x}} \wedge \underset{\sim}{z} = \pi(\underset{\sim}{\tilde{x}})),$$

a contradiction.

To get π, note that x is not symmetric, so that there is $\rho \in \mathscr{G}$ such that $\rho(x) \neq x$ and $\rho \in H \cap \operatorname{fix}(E)$, where E is the projection of e into A. Since $|p| < \kappa$, there exists $\gamma < \kappa$ such that $(a, \xi) \notin \operatorname{dom}(p)$ and $(a, \xi) \notin e$ for any $a \in A$, $\xi \geqslant \gamma$. Thus we define $\pi \in [\rho]$ as follows:

(6.12) (i) if $a \in E$, then $\pi(a, \xi) = (a, \xi)$ for all $\xi < \kappa$;
(ii) if $a \notin E$, then $\pi(a, \xi) = (\rho a, \gamma + \xi)$
and $\pi(a, \gamma + \xi) = (\rho a, \xi)$ for $\xi < \gamma$
and $\pi(a, \xi) = (\rho a, \xi)$ for $\xi \geqslant 2\gamma$.

It follows that $\pi \in \bar{H} \cap \text{fix}(e)$ and that p and πp are compatible.

The following lemma will complete the proof of Theorem 6.1.

LEMMA 6.5. $(\mathscr{P}^\alpha(A)^{\mathscr{V}})^{\sim} = \mathscr{P}^\alpha(\tilde{A})^{\mathscr{N}}$.

PROOF. It suffices to show that the right-hand side is included in the left-hand side. The proof is by induction; we show that if $x \in \mathscr{P}^\alpha(A)^{\mathscr{V}}$ and $y \subseteq x$, $y \in \mathscr{N}$, then there exists $z \in \mathscr{V}$ such that $y = \tilde{z}$. Let $\underline{y}$ be a name for y. Note that every descending sequence of less than κ conditions has a lower bound. Since $|x| < \kappa$, we may use the same argument as in the proof of Lemma 5.12 (or Problem 5.5), to obtain $p \in G$ which decides $\underline{\tilde{t}} \in \underline{y}$ for all $t \in x$. It follows that $y = \tilde{z}$, where $z = \{t \in x : p \Vdash \underline{\tilde{t}} \in \underline{y}\}$. By Lemma 6.4, since $\tilde{z} \in \mathscr{N}$, we have $z \in \mathscr{V}$.

6.2. Refinements of the First Embedding Theorem

The First Embedding Theorem enables us to embed a segment of a given permutation model into a symmetric model. Thus the statements whose validity depends only on a segment of the universe are true in the permutation model if and only if they are true in the corresponding symmetric model. (The reader will find a precise formulation and examples in the problem section). This makes it possible, in many cases, to reduce the problem of consistency in ZF to the construction of a permutation model. Not every property of the permutation model $\mathscr{V}$ holds in the symmetric model $\mathscr{N}$ – naturally, because $\mathscr{V}$ is a model of ZFA while $\mathscr{N}$ is a model of ZF. (We shall give some specific counterexamples in Chapter 9.) We can look at the First Embedding Theorem as a first approximation of the permutation model by a symmetric model. The method described in Section 6.1 can be refined to transfer additional properties of $\mathscr{V}$ into $\mathscr{N}$.

In the proof of the independence of the Axiom of Choice from the Ordering Principle, the key property of the model (in both cases) was the existence of a least support for every element of the model. This property plays an important role in the construction of both permutation and symmetric models; it is our intention to show that this property can be preserved in the transfer from $\mathscr{V}$ to $\mathscr{N}$.

Let $\mathscr{V}$ be a permutation model determined by a group $\mathscr{G}$ of permutations of A, and the ideal I of the finite subsets of A. We say that $\mathscr{V}$ is a *support model* if the group $\mathscr{G}$ has the following property:

(6.13) If E_1, E_2 are finite subsets of A and if $\pi \in \text{fix}(E_1 \cap E_2)$, then there exists a finite number of permutations $\rho_1, \sigma_1, \rho_2, \sigma_2, \ldots, \rho_n, \sigma_n$ such that $\rho_i \in \text{fix}(E_1)$ and $\sigma_i \in \text{fix}(E_2)$ $(i = 1, \ldots, n)$ and $\pi = \rho_1 \cdot \sigma_2 \cdot \ldots \cdot \rho_n \cdot \sigma_n$.

If $\mathscr{G}$ has the property (6.13), then every $x \in \mathscr{V}$ has a least support, and the relation 'E is the least support of x' is in the model $\mathscr{V}$; see Section 4.5 for the details.

If we embed a support model in a symmetric model by the method described in Section 6.1, then obviously every name $\underline{x} \in \mathscr{M}^B$ is symmetric if and only if there exists a pair (e, E) of finite sets, $e \subseteq A \times \kappa$ and $E \subseteq A$ such that $\text{sym}_{\bar{\mathscr{G}}}(\underline{x}) \supseteq \text{fix}(e, E)$, where

(6.14) $\text{fix}(e, E) = \{\pi \in \overline{\mathscr{G}} : \pi \in \text{fix}(e) \text{ and } \pi \in [\rho] \text{ for some } \rho \in \text{fix}(E)\}$.

We can – and always will – assume that E includes the projection of e into A. We may say that (e, E) is a *support* of $\underline{x}$. Actually, we may (as we did in Section 5.5) identify $e = \{(a_1, \xi_1), \ldots, (a_k, \xi_k)\}$ with the canonical name $\underline{e}$ of the set

(6.15) $$e = \{x_{a_1\xi_1}, \ldots, x_{a_k\xi_k}\},$$

and similarly, $E = \{a_1, \ldots, a_k\}$ with the canonical name $\underline{E}$ of the finite set

(6.16) $$E = \{\tilde{a}_1, \ldots, \tilde{a}_k\}.$$

The class

(6.17) $$\underline{\Delta} = \{((\underline{e}, \underline{E})^B, \underline{x})^B : (\underline{e}, \underline{E}) \text{ is a support of } \underline{x}\},$$

is symmetric and its interpretation defines a relation Δ in $\mathscr{N}$:

(6.18) $$(e, E) \text{ is a } \textit{support} \text{ of } x \leftrightarrow ((e, E), x) \in \Delta.$$

We say that $\mathscr{N}$ is a *support model* if every $x \in \mathscr{N}$ has a least support, i.e., a support (e, E) with the property that whenever (e', E') is also a support of x then $e \subseteq e'$ and $E \subseteq E'$.

THEOREM 6.6 (Support Theorem). *If $\mathscr{V}$ is a support model (of* ZFA) *and if we use the method of Section* 5.1 *to embed $\mathscr{V}$ in a symmetric model $\mathscr{N}$, then $\mathscr{N}$ is a support model (of* ZF).

PROOF. We shall more or less follow the proof of Lemma 5.23, and verify that it goes through in the present situation. Let $x \in \mathscr{N}$ and let both (e_1, E_1) and (e_2, E_2) be supports of x; we want to show that $(e_1 \cap e_2, E_1 \cap E_2)$

is a support of x. It suffices to consider three special cases: (1) $E_1 = E_2$, (2) $e_1 = e_2$, and (3) $e_1 \subseteq e_2$ and $E_2 \subseteq E_1$. The general case is deduced as follows:

(e_1, E_1) and (e_2, E_2) are supports,

$(e_1, E_1 \cup E_2)$ and $(e_2, E_1 \cup E_2)$ are supports,

$(e_1 \cap e_2, E_1 \cup E_2)$ is a support, (use case (1))

$(e_1 \cap e_2, E_1)$ and $(e_1 \cap e_2, E_2)$ are supports, (use case (3))

$(e_1 \cap e_2, E_1 \cap E_2)$ is a support. (use case (2))

We are proving the analogue of Lemma 5.23. Let $p \Vdash \underline{x} = \underline{y}$. First note that the analogue of Lemma 5.24 holds:

If $(\underline{e}, \underline{E})$ *is a support for each of the* $\underline{x}_1, \ldots, \underline{x}_n \in \mathrm{HS}$, *then for each condition* p,

(6.19) *if* $p \Vdash \phi(\underline{x}_1, \ldots, \underline{x}_n)$, *then* $p : \underline{e} \Vdash \phi(\underline{x}_1, \ldots, \underline{x}_n)$.

(Note that (6.19) does not depend on $\underline{E}$, only on $\underline{e}$.) The proof goes exactly as in Lemma 5.24: For every $q \supseteq p : \underline{e}$, there exists a permutation $\pi \in \mathrm{fix}\,(\underline{e}, \underline{E})$ such that p and πq are compatible.

Since $\mathscr{V}$ is a support model and because our permutations act inside the blocks arbitrarily, we can easily conclude that every symmetric *name* has a least support $(\underline{e}, \underline{E})$. We also notice that if $(\underline{e}, \underline{E})$ is a support of $\underline{w}$, if ρ_1, ρ_2 are permutations of A, and if $\pi_1 \in [\rho_1]$, $\pi_2 \in [\rho_2]$, then

(6.20) if $\rho_1|\underline{E} = \rho_2|\underline{E}$ and $\pi_1|\underline{e} = \pi_2|\underline{e}$, then $\pi_1(\underline{w}) = \pi_2(\underline{w})$

(the analogue of (5.24)).

Let $p \Vdash \underline{x} = \underline{y}$ and let $(\underline{e}_1, \underline{E}_1)$ and $(\underline{e}_2, \underline{E}_2)$ be supports for $\underline{x}$ and $\underline{y}$, respectively. By (6.19), we may assume that

(6.21) $$s(p) \subseteq \underline{e}_1 \cup \underline{e}_2.$$

Case 1: $\underline{E}_1 = \underline{E}_2$. The proof is virtually the same as the proof of Lemma 5.23. Of course, we consider only those permutations which preserve each $a \in E_1$.

Case 2: $\underline{e}_1 = \underline{e}_2$. In this case, we use the property (6.13) of $\mathscr{G}$. If $\pi \in \mathrm{fix}\,(\underline{e}_1, \underline{E}_1 \cap E_2)$, then we prove that $p \Vdash \underline{x} = \pi\underline{x}$: there are $\rho_1, \sigma_1, \ldots, \rho_k, \sigma_k$ such that $\rho_i \in \mathrm{fix}\,(\underline{e}_1, \underline{E}_1)$, $\sigma_i \in \mathrm{fix}\,(\underline{e}_1, \underline{E}_2)$, $i = 1, \ldots, k$, and by (6.21), $\rho_i p = \sigma_i p = \pi p = p$; thus we have:

$$(6.22)\qquad \begin{aligned} &p \Vdash \sigma_k \underline{x} = \sigma_k \underline{y} = \underline{y} = \underline{x},\\ &p \Vdash \rho_k \sigma_k \underline{x} = \rho_k \underline{x} = \underline{x},\\ &\ \vdots\\ &p \Vdash \pi \underline{x} = \rho_1 \dots \sigma_k \underline{x} = \rho_1 \underline{x} = \underline{x}. \end{aligned}$$

Now we can simply define $\underline{z} \in \mathcal{M}^B$ as follows: for $\underline{w} \in \mathrm{HS}$, $\rho(\underline{w}) < \rho(\underline{x})$, let

$$(6.23)\qquad \underline{z}(\underline{w}) = [\![\underline{w} \in \underline{x}]\!] \cdot p.$$

Obviously, $p \Vdash \underline{z} = \underline{x}$, and $(\underline{e}_1, \underline{E}_1 \cap \underline{E}_2)$ is a support for $\underline{z}$ because if $\pi \in \mathrm{fix}\,(\underline{e}_1, \underline{E}_1 \cap \underline{E}_2)$, then $\pi p = p$ and

$$(6.24)\qquad \underline{z}(\pi \underline{w}) = [\![\pi \underline{w} \in \underline{x}]\!] \cdot p = [\![\pi \underline{w} \in \pi \underline{x}]\!] \cdot \pi p = \pi(\underline{z}(\underline{w})) = (\pi \underline{z})(\pi \underline{w}),$$

and hence $\pi \underline{z} = \underline{z}$.

Case 3: $\underline{e}_1 \subseteq \underline{e}_2$ and $E_2 \subseteq E_1$. For simplicity, let $\underline{E}_1 = \{a, b\}$, $\underline{E}_2 = \{a\}$, $\underline{e}_2 = \{(a, \xi), (a, \eta)\}$, $\underline{e}_1 = \{(a, \xi)\}$. Following the proof of Lemma 5.23, we define

$$(6.25)\qquad \underline{z}(\underline{w}) = [\![\underline{w} \in \underline{x}]\!] : (s(\underline{w}) \cup \underline{e}_1),$$

where $s(\underline{w})$ is the $\underline{e}$ in the least support $(\underline{e}, \underline{E})$ of $\underline{w}$, and this we do for one chosen $\underline{w}$ in each $\{\pi \underline{w} : \pi \in \mathrm{fix}\,(\underline{e}_1, \underline{E}_2)\}$, and the $\underline{w}$ is chosen such that $s(\underline{w})$ is disjoint from $\underline{e}_2$ and from b. When done this way, $\underline{z}(\pi \underline{w})$ can be unambiguously defined as $\pi(\underline{z}(\underline{w}))$ and $\mathrm{sym}\,(\underline{z}) \supseteq \mathrm{fix}\,(\underline{e}_1, \underline{E}_2)$.

To show that $p \Vdash \underline{x} = \underline{z}$, we proceed as in Lemma 5.23. At the point where we specify the permutation π (see (5.35)), we notice that π may be taken as a composition $\pi_1 \pi_2$ of two permutations π_1 and π_2 such that

$$(6.26)\qquad \begin{aligned} &\pi_1 \in \mathrm{fix}\,(e_1, E_1),\\ &\pi_2(a, \xi) = (a, \xi) \text{ for all } \xi < \kappa. \end{aligned}$$

We have $\pi_1 \underline{x} = \underline{x}$, $\pi_2 \underline{y} = \underline{y}$, and since $s(p) \subseteq e_2$, we also have $\pi_2 p = p$ and $\pi p = \pi_1 p$. The proof is completed by showing the analogue of (5.36). To see that $\pi p \cup p \Vdash \pi x = \underline{x}$, we use

$$(6.27)\qquad \begin{aligned} \pi p &= \pi_1 p \Vdash \underline{x} = \pi_1 \underline{x} = \pi_1 \underline{y},\\ \pi p &= \pi_1 \pi_2 p \Vdash \pi \underline{x} = \pi_1 \pi_2 \underline{y} = \pi_1 \underline{y}. \end{aligned}$$

In Section 5.5, when we proved that the Ordering Principle holds in the basic Cohen model $\mathcal{N}$, we used the existence of a least support for each $x \in \mathcal{N}$ to construct a one-to-one function F in $\mathcal{N}$ with values (e, α), where $\alpha \in \mathrm{On}$ and e is a finite set of reals. The support model $\mathcal{N}$ obtained from $\mathcal{V}$ by the method of Section 6.1 enjoys a similar property.

THEOREM 6.7. *If $\mathscr{V}$ is a support model (of ZFA) and if we use the method of Section 6.1 to embed $\mathscr{V}$ in a support model $\mathscr{N}$ (of ZF), then there is a one-to-one function F in $\mathscr{N}$ which maps $\mathscr{N}$ into $\mathscr{P}(\kappa) \times \mathscr{V}$.*

(Note that the family of all subsets of κ is linearly ordered by the lexicographical ordering. We shall refer to it as the *natural linear ordering* of $\mathscr{P}(\kappa)$.)

PROOF. We follow the proof that the Ordering Principle holds in the basic Cohen model and use the Support Theorem. An *assignment* is a finite function

$$t(\underset{\sim}{x}_1) = i(\pi \underset{\sim}{x}_1), \ \ldots, \ t(\underset{\sim}{x}_k) = i(\pi \underset{\sim}{x}_k),$$
$$t(\underset{\sim}{a}_1) = i(\pi \underset{\sim}{a}_1), \ \ldots, \ t(\underset{\sim}{a}_n) = i(\pi \underset{\sim}{a}_n),$$

where $\underset{\sim}{a}_1, \ldots, \underset{\sim}{a}_n$ are canonical names for elements of $\tilde{A}$ and $\underset{\sim}{x}_1, \ldots, \underset{\sim}{x}_k$ are canonical names for elements of elements of $\tilde{A}$ (see (6.1)–(6.3)). We let

$$e(t, \underset{\sim}{x}) = i(\pi \underset{\sim}{x})$$

for every $\underset{\sim}{x} \in \mathrm{HS}$ and an assignment t such that the domain of t includes the support of $\underset{\sim}{x}$. The function e is in $\mathscr{N}$. We construct F as follows: Let $x \in \mathscr{N}$. Let (e, E) be the least support of x, and let $e = \{x_1, \ldots, x_k\}$, where $x_1 < \ldots < x_k$ in the natural linear ordering. There exist $\underset{\sim}{y}_1, \ldots, \underset{\sim}{y}_k$, $\underset{\sim}{a}_1, \ldots, \underset{\sim}{a}_n, \underset{\sim}{y} \in \mathrm{HS}$, and an assignment t such that

$$\text{(6.28)} \qquad \begin{gathered} \operatorname{dom}(t) = \{\underset{\sim}{y}_1, \ldots, \underset{\sim}{y}_k, \underset{\sim}{a}_1, \ldots, \underset{\sim}{a}_n\}, \qquad \operatorname{rng}(t) = e \cup E, \\ t(\underset{\sim}{y}_1) = x_1, \ \ldots, \ t(\underset{\sim}{y}_k) = x_k. \end{gathered}$$

Let $\underset{\sim}{y}_1, \ldots, \underset{\sim}{y}_k, \underset{\sim}{a}_1, \ldots, \underset{\sim}{a}_n, \underset{\sim}{y}$ be least (in the well-ordering of $\mathscr{M}^{\mathbf{B}}$) such that (6.28) holds for some assignment t. Obviously, the set S' of all such assignments is finite (of size at most the number of permutations of E), and since every such t has fixed values on $\underset{\sim}{y}_1, \ldots, \underset{\sim}{y}_k$, the set S' can be defined from a finite set S of finite functions from $\{\underset{\sim}{a}_1, \ldots, \underset{\sim}{a}_n\}$ to E. The set S is in $\mathscr{V}$ and since $\mathscr{M}^B \subseteq \mathscr{V}$, we can assign to each $x \in \mathscr{N}$ a pair (e, v), where e is the first part of the least support of x and $v = (\langle \underset{\sim}{y}_1, \ldots, \underset{\sim}{y}_k, \underset{\sim}{a}_1, \ldots, \underset{\sim}{a}_n, \underset{\sim}{y} \rangle, S)$ where $v \in \mathscr{V}$. This gives a one-to-one function from $\mathscr{N}$ into $I \times \mathscr{V}$, where I is the family of all finite subsets of $\mathscr{P}(\kappa)$. Now we use the fact that there is a one-to-one correspondence between I and $\mathscr{P}(\kappa)$, and thus we obtain a one-to-one function which maps $\mathscr{N}$ into $\mathscr{P}(\kappa) \times \mathscr{V}$.

Now we shall deal with another refinement of the First Embedding Theorem. In Section 6.1, we started with a permutation model $\mathscr{V}$ given by $\mathscr{G}$ and $\mathscr{F}$, and a prescribed ordinal α and constructed a symmetric model $\mathscr{N}$ such that $(\mathscr{P}^\alpha(A))^\mathscr{V}$ is isomorphic to $(\mathscr{P}^\alpha(\tilde{A}))^\mathscr{N}$ for some $\tilde{A} \in \mathscr{N}$. This was done by constructing a generic model $\mathcal{M}[G]$ and then specifying a group $\overline{\mathcal{G}}$ of automorphisms of B and a normal filter $\overline{\mathcal{F}}$ on $\overline{\mathcal{G}}$. The filter $\overline{\mathcal{F}}$ was defined in (6.9), and it is actually the least filter which can simulate $\mathcal{F}$ and make sure that each $x_{a\xi}$ is in $\mathcal{N}$.

However, in the proof of Theorem 6.1, we used only some specific properties of the filter $\overline{\mathcal{F}}$. It turns out that these properties can be satisfied by other, richer filters on $\overline{\mathcal{G}}$, which give us more control over the model $\mathcal{N}$, while still preserving the basic requirement that $\mathcal{V}$ embeds in $\mathcal{N}$.

Let $(P, <)$ be the forcing conditions used in the proof of Theorem 6.1 and let $\overline{\mathcal{G}}$ be as defined in (6.6). For every $\pi \in \overline{\mathcal{G}}$, let ρ_π be the unique $\rho \in \mathcal{G}$ such that $\pi \in [\rho]$. Let $\overline{\mathcal{F}}$ be a normal filter on $\overline{\mathcal{G}}$ generated by a set $\mathcal{K}$ of subgroups of $\overline{\mathcal{G}}$. We say that $\overline{\mathcal{F}}$ is *feasible* if the following two conditions are satisfied:

(6.29) If $H \in \mathcal{F}$, then $\bar{H} \in \overline{\mathcal{F}}$, and conversely, if $H \in \overline{\mathcal{F}}$, then $\{\rho_\pi : \pi \in H\} \in \mathcal{F}$

and

(6.30) If $X \subseteq A \times \kappa$ and $|X| < \kappa$, and if $K \in \mathcal{K}$ and $\rho \in \mathcal{G}$ are such that $K \cap [\rho] \neq 0$, then there exists a permutation $\pi \in K \cap [\rho]$ such that for each $x \in X$, either $\pi x = x$ or $\pi x \notin X$.

Notice that if the condition (6.30) is satisfied, then for every condition $p \in P$ and every K and ρ as in (6.30), there exists $\pi \in K \cap [\rho]$ such that p and πp are compatible. If one goes through the proof of the First Embedding Theorem, one can see that the conditions (6.29) and (6.30) are exactly what we need in the proof of Lemma 6.4. Thus we have the following theorem:

THEOREM 6.8 (Second Embedding Theorem). *Under the assumptions of Theorem* 6.1, *if B and $\overline{\mathcal{G}}$ are constructed as in the proof of Theorem* 6.1, *and if $\overline{\mathcal{F}}$ is a feasible normal filter on $\overline{\mathcal{G}}$, then the corresponding symmetric model satisfies the conclusion of Theorem* 6.1.

The filter constructed in (6.9) is an example of a feasible filter. For another example, see Problem 12. We will have use for it in Chapter 8.

6.3. Problems

1. Let ϕ be a formula of the form $\exists X\, \psi(X, v)$, where the only quantifiers we allow in ψ are $\exists u \in \mathcal{P}^v(X)$ and $\forall u \in \mathcal{P}^v(X)$. Let $\mathcal{V}$ be a permutation model such that $\mathcal{V} \models \exists X\, \phi(X)$. Then there exists a symmetric model $\mathcal{N}$ of ZF such that $\mathcal{N} \models \exists X\, \phi(X)$.

[*Hint*: Let $X \in \mathscr{V}$ be such that $\mathscr{V} \models \psi(X, v)$; let α be such that $\mathscr{P}^{v}(X) \subseteq \mathscr{P}^{\alpha}(A)$. By Theorem 6.1, there exists $\mathscr{N}$ such that $\mathscr{P}^{\alpha}(A)^{\mathscr{U}} \approx \mathscr{P}^{\alpha}(\tilde{A})^{\mathscr{N}}$. Thus $\mathscr{N} \models \psi(\tilde{X}, v)$.]

In Problems 2–10, show that the formula $\phi(X)$ is of the type described in Problem 1 and conclude, using the results from Chapter 4, that the existence of such X is consistent with the axioms of ZF. (The statement in Problem 10 is shown to be consistent with ZFA in Chapter 10.)

2. X cannot be well-ordered.

3. X cannot be linearly ordered.

4. X is a countable set of pairs without a choice function.

5. X is amorphous.

6. X is infinite and $\mathscr{P}(X)$ is Dedekind finite.

7. X is infinite and D-finite and $\mathscr{P}(X)$ is D-infinite.

8. X is infinite and T-finite.

9. X is T-infinite and D-finite.

10. X_1 is a vector space over the field X_2 and X_1 has no basis.

11. The Selection Principle is independent from the Ordering Principle in ZF.

[*Hint*: Express the Selection Principle by a formula of the type described in Problem 1 and embed the linearly ordered Mostowski model $\mathscr{V}$ in a support model $\mathscr{N}$ of ZF. By the Embedding Theorem, the Selection Principle fails in $\mathscr{N}$. By Theorem 6.7, $\mathscr{N}$ has a one-to-one function F into $\mathscr{P}(\kappa) \times \mathscr{V}$, and both $\mathscr{P}(\kappa)$ and $\mathscr{V}$ are linearly ordered. Thus $\mathscr{N}$ is linearly ordered].

12. Let $\aleph_\alpha$ be a cardinal and let A be the set of all atoms, $|A| \geqslant \aleph_\alpha$. Let $\mathscr{G}$ be a group of permutations of A and let $\mathscr{F}$ be the normal filter determined by the ideal of sets of cardinality less than $\aleph_\alpha$ (i.e., $\mathscr{F}$ is generated by $\{\text{fix}(E) : |E| < \aleph_\alpha\}$). Let $\mathscr{M}^B$ be constructed as in Theorem 6.1 and let $\overline{\mathscr{G}}$ be as in (6.6). Let $\overline{\mathscr{F}}$ be the filter on $\overline{\mathscr{G}}$ generated by $\{\bar{H} : H \in \mathscr{F}\} \cup \{\text{fix}(e) : e \subseteq A \times \kappa \text{ and } |e| < \aleph_\alpha\}$. Then $\overline{\mathscr{F}}$ is feasible.

The following result is mentioned here because its proof is closely related to the Embedding Theorem and its application gives counterexamples to the Axiom of Choice. The proof requires some facts from model theory. A theory T is $\aleph_0$-*categorical* if T has a unique (up to isomorphism) countable model. (The most popular example is the theory of linear order.)

**13. If A is the countable model of an $\aleph_0$-categorical theory T (in $\mathcal{M}$), then there is a symmetric extension $\mathcal{N}$ of the ground model $\mathcal{M}$ and $A' \in \mathcal{N}$ isomorphic to A such that the only subsets of A' in $\mathcal{N}$ are those definable over A' (in the language of T) with parameters from A'.

This can be proved by using the automorphisms of A as permutations and finite subsets as supports to construct a permutation model, and then to use the Embedding Theorem to get a symmetric model of ZF. One uses facts such as: T has only finitely many n-types for each n, etc..

Here are several applications: The statements in Problems 14–16 are consistent with ZF:

14. There is an amorphous set.

[*Hint*: Let T be the theory with no axioms; T is $\aleph_0$-categorical. Every subset of A' is either finite or a complement of a finite set.]

15. There is a linearly ordered set whose only subsets are finite unions of intervals and points.

[*Hint*: Let T be the $\aleph_0$-categorical theory of linear order.]

*16. There is a vector space without a basis.

[*Hint*: Let T be the $\aleph_0$-categorical theory of infinite-dimensional vector spaces over a two element field. Let $\phi(x, \vec{p})$ define a basis B in A; let $x, y, z \in B$ be such that neither of them is a linear combination of the p's. There is an automorphism π of A which leaves $\vec{p}$ fixed and $\pi(x) = y+z$. Then $y+z \in B$, a contradiction.]

*17. (For model theorists.) The A' from Problem 13 is D-finite.

[*Hint*: If A is the countable model of an $\aleph_0$-categorical theory, then no infinite subset of A has a well-ordering definable over A with parameters from A.]

6.4. Historical remarks

The two Embedding Theorems and their applications (Problems 1–10) are due to Jech and Sochor [1966a, b]. The Support Theorem and the example in Problem 11 are due to Pincus [1969, 1971, 1972a] (Problem 11 is independently due to Felgner [1971b]). The theorem in Problem 13 and the subsequent applications are due to Plotkin [1969].

CHAPTER 7

MODELS WITH FINITE SUPPORTS

7.1. Independence of the Axiom of Choice from the Prime Ideal Theorem

In this chapter, we present several independence results which are obtained by constructing a permutation or a symmetric model with the filter given by the ideal of finite sets. In all the examples, except the first one concerning the Prime Ideal Theorem, it suffices to deal with permutation models, and the Embedding Theorem and the Support Theorem provide an easy transfer to set theory without atoms.

In this section we prove:

THEOREM 7.1. *The Axiom of Choice is independent from the Prime Ideal Theorem.*

Theorem 7.1 is true for set theory ZF (without atoms). There is a symmetric model $\mathcal{N}$ of ZF in which the Axiom of Choice fails while the Prime Ideal Theorem holds. Actually, this is true in the basic Cohen model $\mathcal{N}$. By the results of Section 5.5, the Axiom of Choice fails in $\mathcal{N}$ and every set can be linearly ordered. Since the Prime Ideal Theorem implies the Ordering Principle, we get a stronger result than in Section 5.5.

We shall not give the proof of the Prime Ideal Theorem in the basic Cohen model. Rather, we show that the Prime Ideal Theorem holds in the Mostowski permutation model (see Section 4.5). Similarly, as the proof of the Ordering Principle in the Cohen model requires more subtle methods than the proof in the Mostowski model, although the idea is similar, the proof of the Prime Ideal Theorem is considerably more complicated than the proof in the Mostowski model. In particular, one apparently cannot avoid the necessity of a heavy use of involved combinatorial properties.

The rest of this section is devoted to the proof that the Prime Ideal Theorem holds in the linearly ordered Mostowski model of ZFA.

We recall that the set A of atoms is ordered like the rationals, that $\mathscr{G}$ is the group of all order-preserving permutations of A and that an element x is symmetric if

$$\operatorname{sym}(x) \supseteq \operatorname{fix}(E)$$

for some finite E; E is a support of x. Every x has a least support.

We further recall that the Prime Ideal Theorem is equivalent to the Consistency Principle:

For every binary mess M there exists a function f which is consistent with M.

(See Section 2.3.) We shall show that if M is a mess in the Mostowski model $\mathscr{V}$, then there is $f \in \mathscr{V}$ which is consistent with M.

Let $M \in \mathscr{V}$ be a binary mess on a set S. Let E_M be the least support of M. For every $x \in S$, let

(7.1) $$[x] = \text{the orbit of } x = \{\pi x : \pi \in \operatorname{fix}(E_M)\}$$

and let $\hat{S} = \{[x] : x \in S\}$. We shall define a binary mess $\hat{M}$ on $\hat{S}$. We let $\hat{M}$ consist of all binary functions h defined on finite subsets of $\hat{M}$ that satisfy the following condition:

(7.2) For every finite subset $P \subseteq \bigcup \{[x] : [x] \in \operatorname{dom} h\}$, the following function t on P is in M:

$$t(x) = h([x]) \quad \text{for every } x \in P.$$

If this is the case, we say that P *admits* h. We have to show that the following condition is satisfied, in order to know that $\hat{M}$ is a mess:

(7.3) For every finite $Q \subseteq \hat{S}$, there exists $h \in \hat{M}$ defined on Q.

Once we know that $\hat{M}$ is a mess, we take any g on $\hat{S}$ consistent with $\hat{M}$ and define

(7.4) $$f(x) = g([x]),$$

for every $x \in S$. The function f is obviously symmetric and hence $f \in \mathscr{V}$ and we are done.

Thus all we have to do is to prove (7.3). We shall distinguish two cases:

7.1.1. *Case I: E_M is empty.*

Let Q be a finite subset of $\hat{S} = \{[x] : x \in S\}$. We are looking for a binary function h on Q which satisfies (7.2). Notice that we have $k = 2^q$ functions to choose from, where $q = |Q|$. Let $\bar{Q} = \{x \in S : [x] \in Q\}$. Fix some $P_0 \subset \bar{Q}$ which has exactly one element in each $[x] \in \bar{Q}$ and notice that

$$\bar{Q} = \bigcup \{\pi P_0 : \pi \in \mathscr{G}\}. \tag{7.5}$$

Let us say that $P \subset \bar{Q}$ is an *n-set* if there are n permutations $\pi_1, ..., \pi_n \in \mathscr{G}$ such that $P = \pi_1 P_0 \cup ... \cup \pi_n P_0$. Since we have only k choices of h and since every finite subset of $\bar{Q}$ is included in an n-set for some n, it is sufficient to show that for any n, there exists an h such that every n-set P admits h.

Let n be fixed. We say that two n-sets P_1 and P_2 have the *same type* if $P_2 = \pi_1(P)$ for some $\pi \in \mathscr{G}$. Certainly, if P_1 and P_2 have the same type and if P_1 admits h, then P_2 admits h. Thus it suffices to prove that there exists an h such that for each type, there is an n-set P of that type which admits h.

First we show that there are only finitely many types of n-sets. Let E_0 be the least support of P_0; let $r = |E_0|$. Let $E_1, ..., E_n, E_1', ..., E_n'$ be subsets of A, each having r elements. We say that $\{E_1, ..., E_n\}$ and $\{E_1', ..., E_n'\}$ have the *same type* if there is $\pi \in \mathscr{G}$ which transforms $\{E_1, ..., E_n\}$ into $\{E_1', ..., E_n'\}$. Notice that there are only finitely many types of these n-tuples; e.g., if $E_1 = \{a, b\}$ and $E_2 = \{c, d\}$, then there are exactly three types: $a < b < c < d$, $a < c < d < b$ and $a < c < b < d$. Let m be the number of types of the n-tuples. For each $E = \pi E_0$, let $P_E = \pi P_0$; similarly, for each $\vec{E} = \{E_1, ..., E_n\} = \{\pi_1 E_0, ..., \pi_n E_0\}$, let $P_{\vec{E}} = \pi_1 E \cup ... \cup \pi_n E$. If $\vec{E}$, $\vec{E}'$ have the same type, then $P_{\vec{E}}$ and $P_{\vec{E}'}$ have the same type. Thus it suffices to find h such that each of the m types of n-tuples admits h.

At this point, we have to resort to combinatorial methods.

LEMMA 7.2. *There exists a finite set U of r-element subsets of A which is so big that whenever we partition U into $k = 2^q$ disjoint sets $U = U_1 \cup ... \cup U_k$, then at least one U_i contains as subsets n-tuples $\{E_1, ..., E_n\}$ of all m types.*

Once we prove Lemma 7.2, we will be done. For consider such a set U and let $P = \bigcup \{P_E : E \in U\}$. Let $t \in M$ be a function defined on P. Each $t|P_E$ corresponds to one of the k possible functions $h_1, ..., h_k$ on Q; this induces a partition of U into k parts. Some U_i contains n-tuples of all types, and so every type admits the function h_i.

To prove Lemma 7.2, we use a popular combinatorial lemma, known as Ramsey's Theorem. If W is a set and r a number, then $[W]^r = \{X \subseteq W: |X| = r\}$. If $[W]^r = Y_1 \cup ... \cup Y_k$ is a partition, then a set $H \subseteq W$ is *homogeneous* (for that partition) if $[H]^r$ is included in some Y_i.

LEMMA 7.3 (Ramsey's Theorem). *Let k and r be given numbers. For every number z, there is a number w such that whenever W is a set having w elements and $[W]^r = Y_1 \cup ... \cup Y_k$ is a partition, then there is a homogeneous set H which has z elements.*

For the proof of Ramsey's Theorem, see Problem 3.

PROOF OF LEMMA 7.2. First notice that whenever H is a set of atoms and $|H| = n \cdot r$, then each of the types of n-tuples is represented by some $\{E_1, ..., E_n\}$ with $E_i \subseteq H$, $i = 1, ..., n$. Let $z = n \cdot r$. Let w be a number satisfying Ramsey's Theorem. Let W be a set of w atoms and let $U = [W]^r$. If $U = U_1 \cup ... \cup U_k$, then by Ramsey's Theorem, there is $H \subset W$ of size $n \cdot r$ such that $[H]^r \subset U_i$ for some i. The set U_i contains as subsets n-tuples of all m types.

This completes the proof of Case I.

7.1.2. *Case II: E_M is nonempty*

Here the combinatorics gets more involved. For simplicity, assume that $E_M = \{a\}$. The proof goes as before, except that we have fix (E_M) instead of $\mathscr{G}$, and E ranges over subsets of A not of size r but of a given *signature* $\vec{r} = (r_1, r_2)$, where r_1 is the number of elements of E_0 less than a, and r_2 the number of elements greater than a. Again, the key is Lemma 7.2, with 'r-element subsets' replaced by 'sets of signature $\vec{r}$'.

As before, all we have to do to prove Lemma 7.2 is to produce a set $H \subset A$ of signature (nr_1, nr_2) such that $[H]^{(r_1,r_2)} \subset U_i$, where $[H]^{(r_1,r_2)} = \{X \subseteq H: X \text{ has signature } (r_1, r_2)\}$. Instead of Ramsey's Theorem, we use the following lemma:

LEMMA 7.4. *Let k, r_1 and r_2 be given numbers. For every number z, there is a number w such that whenever W is a set having w elements and $[W]^{r_1} \times [W]^{r_2} = Y_1 \cup ... \cup Y_k$ is a partition, then there exist Z_1 and Z_2, each having z elements, such that $[Z_1]^{r_1} \times [Z_2]^{r_2} \subseteq Y_i$ for some i.*

For the proof of this lemma, a generalization of Ramsey's Theorem, see Problems 4 and 5. The case when E_M has more than one element is handled similarly.

7.2. Independence of the Prime Ideal Theorem from the Ordering Principle

The Prime Ideal Theorem implies that every set can be linearly ordered, as shown in Section 2.3, where the Prime Ideal Theorem is used to prove the Order Extension Principle which says that every partial ordering can be extended to a linear ordering. In this section, we show that the Ordering Principle is weaker than the Prime Ideal Theorem by exhibiting a model (first a permutation model and then a symmetric model) in which every set can be linearly ordered but in which there exists a partial ordering which cannot be extended to a linear ordering; consequently, the Prime Ideal Theorem fails in the model. The partial ordering that will be used in the model is the universal homogeneous partial ordering of a countable set. Before we construct the model, we better say a few words about the universal homogeneous partial ordering.

A countable partially ordered set $(P, <)$ is *universal* if every finite partially ordered set $(E, \prec)$ can be embedded in $(P, <)$. $(P, <)$ is *homogeneous* if whenever E_1 and E_2 are finite subsets of P, and i is an isomorphism of $(E_1, <)$ and $(E_2, <)$, then i can be extended to an automorphism of $(P, <)$.

Notice that if we consider linear ordering rather than partial ordering, then the rationals are the unique (up to isomorphism) countable universal homogeneous linearly ordered set. The proof of uniqueness – the well-known zig-zag argument – uses a property somewhat stronger than universality, which, however, follows from universality and homogeneity. Using the following lemma, we can similarly show that the countable universal homogeneous partial ordering (if it exists) is unique (Problem 7). Moreover, it follows easily from Lemma 7.5, that every *countable* $(E, \prec)$ can be embedded in $(P, <)$.

LEMMA 7.5. *Let $(P, <)$ be a countable universal homogeneous partially ordered set. If $(E, \prec)$ is a finite partially ordered set, $E_0 \subseteq E$, and if e_0 is an embedding of $(E_0, \prec)$ into $(P, <)$, then there is an embedding e of $(E, \prec)$ into $(P, <)$ which extends e_0.*

PROOF. We may assume that $E_0 \subseteq P$ and that e_0 is the identity on E_0. The set $(E, \prec)$ can be embedded in $(P, <)$; let i be an embedding. The mapping $i|E_0$ can be extended to an automorphism π of $(P, <)$. Then $e = \pi^{-1} \cdot i$ (i followed by π^{-1}) is an embedding of $(E, \prec)$ in $(P, <)$ which is the identity on E_0.

We shall show that a countable universal homogeneous partial ordering exists. We shall stress which properties of partial order we use, so that the reader can see that the existence of universal homogeneous partial orders is a special case of a much more general theorem.

Let $P_0 \subseteq P_1 \subseteq \ldots \subseteq P_n \subseteq \ldots$ be a sequence of finite sets, each partially ordered such that the ordering of P_n is the restriction of the ordering of P_{n+1}. Then $P = \bigcup_{n=0}^{\infty} P_n$ is the *direct sum* of the partially ordered sets P_n, $n = 0, 1, \ldots$.

Let A and B be two finite partially ordered sets and let $A_0 \subseteq A$ and $B_0 \subseteq B$ be such that A_0 is isomorphic to B_0. Then there is a partially ordered set $A' \cup B'$ such that A' is isomorphic to A, B' is isomorphic to B, and $A' \cap B'$ is isomorphic to B_0. To see this we may assume that $A \cap B = A_0 = B_0$ and the orderings of A and B agree on B_0. All we have to do is to define the order between the elements of A and the elements of B. The result is an *amalgamated sum* of A and B (w.r.t. A_0, B_0).

The above two properties are used to construct a universal homogeneous partial ordering.

LEMMA 7.6. *A countable universal homogeneous partially ordered set exists.*

PROOF. Let $P = \{p_0, p_1, ..., p_n, ...\}$ be a countable set; we construct a universal homogeneous partial order on P by constructing finite $P_0 \subseteq P_1 \subseteq ... \subseteq P_n \subseteq ...$ such that P is the direct sum of P_n's. Let $E_0, E_1, ..., E_n, ...$ be a sequence of finite partially ordered sets such that every finite partially ordered set is isomorphic to some E_n. Let $i_0, i_1, ..., i_n, ...$ be an enumeration of all i's such that i is an isomorphism of two finite partially ordered sets A_n and B_n, where both $A_n \subseteq P$ and $B_n \subseteq P$. (We may assume that $A_n \cup B_n \subseteq \{p_0, ..., p_n\}$.) We construct $(P_n, <)$, $n = 0, 1, ...$, so that each P_n is $\{p_0, ..., p_k\}$ for some k, and the ordering $<$ on P_n is the restriction of the ordering $<$ on P_{n+1}. If $P_{2n} = \{p_0, ..., p_k\}$ and $|E_n| = m$, then $P_{n+1} = P_n \cup \{p_{k+1}, ..., p_{k+m}\}$, with the added elements ordered isomorphically to E_n. If $P_{2n+1} = \{p_0, ..., p_k\}$, then if the orderings of A_n and B_n agree with the ordering of P_{2n+1}, consider an amalgamated sum of P_{2n+1} and P_{2n+1} with respect to A_n and B_n. We can assume that the sum is $P_{2n+1} \cup Q$, where $Q = B_n \cup \{p_{k+1}, ..., p_{k+m}\}$. This gives us an extension of i_n to the set P_{2n+1}.

The direct sum $P = \bigcup_{n=0}^{\infty} P_n$ is universal and homogeneous: Every finite partially ordered set is embedded in P via one of the E_n's. And if i is an isomorphism of two finite subsets of P, then $i = i_{n_0}$ for some n_0 and this extends to an isomorphism j with domain P_{2n_0+1}; then $j = i_{n_1}$ for some n_1, etc.; in this way, we get an extension π of i which is defined everywhere on P (of course, we have to zig-zag between domains and ranges to make sure that the range of π is P).

The reader remembers from Chapter 4 that the rationals also enjoy the following property:

LEMMA 7.7. *Let $(P, <)$ be a countable universal homogeneous partially ordered set and let $\mathscr{G}$ be the group of all automorphisms of $(P, <)$. If E_1 and E_2 are finite subsets of P and if $\pi \in \operatorname{fix}(E_1 \cap E_2)$, then there are $\pi_1, \rho_1, ..., \pi_n, \rho_n$ such that $\pi_i \in \operatorname{fix}(E_1)$ and $\rho_i \in \operatorname{fix}(E_2)$ and $\pi = \rho_n \cdot \pi_n \cdot ... \cdot \rho_1 \cdot \pi_1$.*

PROOF. First we prove the special case when $E_1 = \{a\}$ and $E_2 = \{b\}$. This case breaks into several subcases according to the constellation of a, b, πa and πb; we will handle two typical subcases and leave the rest to the reader. First, let $a < b < \pi a < \pi b$. We let π_1 be an automorphism which extends the isomorphism $i(a) = a$, $i(b) = \pi(b)$, and ρ_1 an automorphism such that $\rho_1(a) = \pi(a)$ and $\rho_2(\pi b) = \pi b$. Although $\rho_1 \cdot \pi_1$ need not be equal to π, we have $\pi = \tilde{\pi} \cdot \rho_1 \cdot \pi_1$, where $\pi = \tilde{\pi} \cdot (\rho_1 \cdot \pi_1)^{-1} \in \text{fix}\,\{a,b\}$. Secondly, let $a < \pi b$ and $b < \pi a$, and otherwise incomparable. We let $a_1 b_1$ be elements incomparable with any of them and with each other and let

$$\pi_1 a = a, \quad \pi_1 b = b_1; \quad \rho_1 a = a_1, \quad \rho_1 b = b, \quad \rho_1 b_1 = \pi b;$$

$$\pi_2 a = a, \quad \pi_2 a_1 = \pi a; \quad \pi_2(\pi b) = \pi b.$$

Next we notice that the same proof works in the case when $E_1 = E \cup \{a\}$ and $E_2 = E \cup \{b\}$ for some E; all the automorphisms involved have to be in fix (E).

The general case when $E_1 \cap E_2 = E$ is proved by induction on $(|E_1 - E|, |E_2 - E|)$. E.g., let $E_2 = E_2' \cup \{a\}$, and by the induction hypothesis: π is a product of automorphisms which leave either E_1 or E_2' fixed. Also by the induction hypothesis, each automorphism which leaves E_2' fixed is a product of automorphisms which leave either $E_2' \cup \{a\}$ or $E_1 \cup E_2'$ fixed.

Now we are almost ready to construct the permutation model. However, we need something a little more general than a universal homogeneous partial ordering. Let us consider structures $(P, <, \prec)$ where $<$ is a partial and $\prec$ a linear ordering of P. An isomorphism of such structures is a one-to-one mapping which preserves both $<$ and $\prec$. We can define a *universal homogeneous structure* in the same way as before, and as before, we can prove the analogues of Lemmas 7.5–7.7 (only the special case of Lemma 7.7 breaks into more subcases, otherwise the proofs are the same).

Let the set A of atoms be countable, and let $<$ and $\prec$ be a partial and a linear ordering such that $(A, <, \prec)$ is a universal homogeneous structure. Let $\mathscr{G}$ be the group of all automorphisms of $(A, <, \prec)$, and let the model $\mathscr{V}$ consist of all hereditarily symmetric elements, where x is symmetric if it has a finite support $E \subseteq A$: sym $(x) \supseteq$ fix (E).

If E_1 and E_2 are supports of $x \in \mathscr{V}$, then $E_1 \cap E_2$ is also a support because every $\pi \in \text{fix}\,(E_1 \cap E_2)$ is a product of automorphisms each leaving either E_1 or E_2 fixed. Thus each $x \in \mathscr{V}$ has a least support and this, as in the Mostowski model (see Lemma 4.6), gives a one-to-one mapping (in $\mathscr{V}$)

of $\mathscr{V}$ into $\mathrm{On} \times I$, where I is the set of all finite subsets of A. Obviously, the linear ordering $\prec$ of A is in $\mathscr{V}$ and so every $x \in \mathscr{V}$ can be linearly ordered in $\mathscr{V}$.

The partial ordering $<$ of A cannot be extended in $\mathscr{V}$ to a linear ordering. To see this, assume that $<^*$ is an extension of $<$ and that $E \subseteq A$ is its support. Let a, b, c, d be elements of A which are all bigger in both orderings $<$ and $\prec$ than all elements of E, and such that:

(i) $a \prec b \prec c \prec d$,

(ii) $a < c$, $d < b$ and otherwise incomparable.

We show that both $a <^* b$ and $a >^* b$ lead to a contradiction. If $a >^* b$, then let $\pi \in \operatorname{fix}(E)$ be such that $\pi a = b$ and $\pi b = c$; since π preserves $<^*$, this gives $b >^* c$, while $a < c$, a contradiction. If $a <^* b$, then we similarly conclude that $b <^* c$ and $c <^* d$. This is a contradiction since $b > d$.

Thus we have a permutation model in which every set can be linearly ordered but not every partial ordering can be extended to a linear ordering. This constitutes the proof of the following theorem (at least in ZFA):

THEOREM 7.8. *The Prime Ideal Theorem is independent from the Ordering Principle.*

PROOF. To establish this independence result for set theory without atoms, we have to produce a symmetric model. To do that, we simply use the First Embedding Theorem (Theorem 6.1). The statement 'A has a partial ordering which cannot be extended to a linear ordering' is of the type described in Problem 6.1, and thus is automatically true in the symmetric model $\mathscr{N}$. To show that $\mathscr{N}$ has linear orderings, we use Theorem 6.7. By Theorem 6.7, $\mathscr{N}$ has a one-to-one function F into $\mathscr{P}(\kappa) \times \mathscr{V}$, and $\mathscr{P}(\kappa)$ is linearly ordered; the permutation model $\mathscr{V}$ is in this case also linearly ordered. Thus we obtain a linear ordering of $\mathscr{N}$.

We conclude this section by the remark that the independence of the Prime Ideal Theorem from the Ordering Principle in the ZF setting can be established directly by the construction of a model similar to the basic Cohen model (see Problem 8).

7.3. Independence of the Ordering Principle from the Axiom of Choice for Finite Sets

In this section, we give an outline of the proof of the following theorem:

THEOREM 7.9. *The Ordering Principle is independent from the Axiom of Choice for families of finite sets.*

We exhibit a model in which every family of finite sets has a choice function but not every set can be linearly ordered. As a matter of fact, the model satisfies a stronger version of the Axiom of Choice than choice from finite sets. The following is true in the model: Every family of well-from finite sets. The following is true in the model:

Every family of well-orderable sets has a choice function.

In Problems 4.14 and 5.22, we gave examples of models, both permutation and symmetric, which satisfy the choice from well-orderable sets. However, both models satisfy the Ordering Principle too (see also Problems 7.9, 7.10). If one analyzes the proof of the choice from well-orderable sets in those models, one can find out that two properties are used: (a) every set has a least support, and (b) every support is assigned one fixed enumeration. Thus we could prove Theorem 7.9 if we had a group of permutations which would satisfy (a) and (b), and if, in addition, the set A could not be linearly ordered. All this is possible in view of the following lemma.

LEMMA 7.10. *There exists a group $\mathscr{G}$ of permutations of a countable set A such that*:

(i). *If E_1, E_2 are finite subsets of A and if $\pi \in \mathscr{G}$ is the identity on $E_1 \cap E_2$, then there exist $\pi_1, \rho_1, \ldots, \pi_n, \rho_n \in \mathscr{G}$ such that $\pi = \rho_n \cdot \pi_n \cdot \ldots \cdot \rho_1 \cdot \pi_1$ and $\pi_i \in \operatorname{fix}(E_1)$, $\rho_i \in \operatorname{fix}(E_2)$, $i = 1, \ldots, n$.*

(ii). *If E is a finite subset of A and if $\pi \in \mathscr{G}$ is such that $\pi''E = E$, then π is the identity on E.*

(iii). *For every finite $E \subseteq A$, there are $a, b, c \in A$ and $\pi, \rho \in \operatorname{fix}(E)$ such that*

$$\pi: a \mapsto b \mapsto c \quad \textit{and} \quad \rho: c \mapsto a \mapsto b.$$

We shall not give a proof of this lemma. The group with these properties cannot be found among frequently used permutation groups. It has to be constructed and the construction is rather involved. Let us only mention that the hard part is to construct a group $\mathscr{G}'$ satisfying (i), (ii) and (iii′), where (iii′) is like (iii) but weakened by the assumption $E = 0$; then one can take $\mathscr{G}$ as a weak direct product of $\aleph_0$ copies of $\mathscr{G}'$. The group $\mathscr{G}'$ is a homomorphic image of the free group with $\aleph_0$ generators, and the kernel is the union of an infinite chain of subgroups which are constructed by recursion. Throughout the recursion, one tries to secure the validity of conditions (i), (ii) and (iii′) in the resulting homomorphic image. For details, we refer the reader to the literature.

With Lemma 7.10 at hand, we shall have no problems in proving Theorem 7.9. First, we construct a permutation model $\mathscr{V}$. We let the set of atoms A

be countable and let $\mathscr{G}$ be a permutation group on A satisfying the conditions 7.10(i)–(iii). The model $\mathscr{V}$ is determined by $\mathscr{G}$ and the ideal of finite subsets of A, i.e., an x is symmetric just in case there is a finite $E \subseteq A$ such that $\text{fix}(E) \subseteq \text{sym}(x)$.

By Lemma 7.10 (i), every $x \in \mathscr{V}$ has a least support. By (ii), if E is the least support of x and $\pi E = E$, then $\pi x = x$. Thus the function which assigns to each $x \in \mathscr{V}$ the pair (E, X), where E is the least support and $X = \{\pi x : \pi \in \mathscr{G}\}$ is the orbit of x, is one-to-one. Also, this function is in $\mathscr{V}$ (i.e., symmetric as a class) and the class of all orbits is in a one-to-one correspondence (also in $\mathscr{V}$) with On. So $\mathscr{V}$ has a one-to-one function F from the universe into $\text{On} \times I$, where I is the set of all finite subsets of A. If $X \in \mathscr{V}$ is well-orderable, then there exists a one-to-one $f \in \mathscr{V}$ from X into the ordinals, and we have $\text{sym}(y) \supseteq \text{fix}(X) = \text{sym}(f)$ for each $y \in X$. Thus the union of the least supports of all elements of X is a finite set E, and the function F restricted to X has values in $\text{On} \times \mathscr{P}(E)$. Now notice that by Lemma 7.10(ii), I has a choice function in $\mathscr{V}$ and this choice function induces a well-ordering of $\mathscr{P}(E)$, canonically for each $E \in I$. Consequently, each $\text{On} \times \mathscr{P}(E)$ has a canonical well-ordering in $\mathscr{V}$, and this enables us to choose an element from each well-orderable X. Thus in $\mathscr{V}$, each family of well-orderable sets has a choice function.

Now we use condition 7.10(iii) to show that A cannot be linearly ordered. Assume that $<$ is a linear ordering of A and that E is a support of $<$. Let $a, b, c \in A$, and let $\pi, \rho \in \mathscr{G}$ be as in condition (iii). Then both $a < b$ and $b < a$ lead to a contradiction. E.g., if $a < b$, then $b = \pi a < \pi b = c$ and so $a < c$; hence $b = \rho a < \rho c = a$, a contradiction.

To establish the result for ZF, we use the First Embedding Theorem and the model $\mathscr{V}$ above to get a symmetric model $\mathscr{N}$ of ZF. The statement 'A cannot be linearly ordered' transfers to $\mathscr{N}$ and we only have to show that in $\mathscr{N}$, every family of well-orderable sets has a choice function. By Theorem 6.6, we can make $\mathscr{N}$ a support model. Then the proof goes as outlined for the basic Cohen model in Problem 5.22.

The first step is to show that $\mathscr{N}$ has a one-to-one function F from the universe into $\text{On} \times I \times J$, where J is a family of finite sets of sets of ordinals and I is the family of finite subsets of the set A. The proof is as in Theorem 6.7. Specifically, the assignment t satisfying (6.28) is unique, since we have a choice function on I and therefore have a fixed enumeration of each finite $E \subseteq A$. Consequently, each $x \in \mathscr{N}$ is uniquely determined by (e, E) and $\underline{y}_1, \ldots, \underline{y}_k, \underline{a}_1, \ldots, \underline{a}_n, \underline{y} \in \mathscr{M}$ and this gives a one-to-one function F of $\mathscr{N}$ into $\text{On} \times I \times J$.

If $X \in \mathcal{N}$ is well-orderable, then since $\mathcal{N}$ is a support model, the argument used in Problem 5.22 shows that there exist finite $E \subseteq \tilde{A}$ and $e \subseteq \bigcup \{a: a \in \tilde{A}\}$ such that whenever $y \in X$ and (E_1, e_1) is the least support of y, then $E_1 \subseteq E$ and $e_1 \subseteq e$. For each well-orderable X, let E and e be minimal with that property. Then the one-to-one function F restricted to X has values in $\mathrm{On} \times \mathscr{P}(E) \times \mathscr{P}(e)$, which has a canonical well-ordering for each E and each e. This gives us the way to pick an element in X for each well-orderable X.

7.4. The Axiom of Choice for Finite Sets

In Chapter 5, we showed that the existence of a choice function for every family of two element sets is unprovable in ZF. In this section, we investigate the relationship between the Axiom of Choice for Finite Sets,

every family of finite sets has a choice function,

and the statements

C_n *every family of n-element sets has a choice function,*

for different $n \in \omega$.

The main results of this section are:

THEOREM 7.11. *The Axiom of Choice for Finite Sets is unprovable even if we assume that* C_n *is true for each n*;

and furthermore, Theorems 7.15 and 7.16, which give a necessary and sufficient condition for C_n being provable from a conjunction of $C_2, \ldots, C_m$.

The independence result will be established by means of a permutation model, but it will be shown how to transfer the results into ZF.

We postpone the proof of Theorem 7.11 and give first two examples to show the relationship between various C_n.

EXAMPLE 7.12. C_2 implies C_4.

PROOF. We are given a family $\mathscr{F}$ of four-element sets and are looking for a choice function f; we assume that we have a choice function g for two-element sets. Let $A \in \mathscr{F}$; we give a rule (using g) for which element of A to choose. There are six two-element subsets of A. For every $a \in A$, let $q(a)$ be the number of all pairs $\{a, b\} \subset A$ such that $g(\{a, b\}) = a$. Obviously $q(a)$ is not the same for all $a \in A$; thus if q is the least such $q(a)$, the set $B = \{a: q(a) = q\}$ has 1, 2 or 3 elements. If B has one element,

$B = \{a\}$, then we let $f(A) = a$. If B has three elements, then $A - B$ has one, $A - B = \{a\}$ and we let $f(A) = a$. If B has two elements, then we let $f(A) = g(B)$.

EXAMPLE 7.13. C_2 does not imply C_3, i.e., C_3 is unprovable from C_2.

First we construct a permutation model $\mathscr{V}$ in which every family of pairs has a choice function and there is a family of triples without a choice function. Assume that the set A of atoms is a countable union $\bigcup_{n=0}^{\infty} T_n$ of triples $T_n = \{a_n, b_n, c_n\}$. Let $\mathscr{G}$ be the group generated by the following permutations π_n of A:

$$\pi_n\colon a_n \mapsto b_n \mapsto c_n \mapsto a_n,$$
$$\pi_n x = x \quad \text{for all } x \notin T_n.$$

($\mathscr{G}$ is the weak direct product of $\aleph_0$ cyclic groups of order 3.) The model $\mathscr{V}$ is determined by $\mathscr{G}$ and the ideal of finite subsets of A. Thus for each $X \in \mathscr{V}$, there exists a finite $E \subseteq A$ such that $\text{sym}(X) \supseteq \text{fix}(E)$.

It is easy to see that the family $\{T_n : n \in \omega\}$ does not have a choice function. For assume that some $f \in \mathscr{V}$ chooses one element from each T_n, and that $E \subseteq A$ is a support for f. Let n be such that $T_n \cap E = 0$; then $\pi_n \in \text{fix}(E)$ so that $\pi_n f = f$ and $\pi_n T_n = T_n$, but $\pi_n x \neq x$ for each $x \in T_n$.

To show that the Axiom of Choice for pairs holds in $\mathscr{V}$, we prove that whenever $S \in \mathscr{V}$ has two elements then:

(7.6) for some $x \in S$, $\text{sym}(x) \supseteq \text{sym}(S)$.

First we show that if $\mathscr{F} \in \mathscr{V}$ and if every $S \in \mathscr{F}$ satisfies (7.6), then $\mathscr{F}$ has a choice function in $\mathscr{V}$. (This holds in every permutation model and we shall use it several times.) $\mathscr{F}$ is a disjoint union of orbits $\{\pi S : \pi \in \text{sym}(\mathscr{F})\}$, $S \in \mathscr{F}$. Let us choose one S in each orbit and for that S, choose $x \in S$ such that $\text{sym}(x) \supseteq \text{sym}(S)$; call this $x = f(S)$. It follows from (7.6) that whenever $\pi(S) = \rho(S)$, then $\pi(x) = \rho(x)$, and so we may define f for all sets in $\mathscr{F}$ by $f(\pi S) = \pi(f(S))$, $\pi \in \text{sym}(\mathscr{F})$. Obviously, $\text{sym}(f) \supseteq \text{sym}(\mathscr{F})$.

To show that (7.6) holds in our model for every S with two elements, let $S = \{x, y\}$ and let $H = \text{sym}(S)$. If we let $K = \text{sym}(x) \cap \text{sym}(S)$, then note that if $\pi, \rho \in H$ then $\pi x = \rho x$ if and only if $\pi \cdot K = \rho \cdot K$. Hence either $K = H$ or $[H : K] = 2$, where $[H : K]$ is the index of K in H, the number of cosets. Thus it suffices to show that $[H : K]$ is not 2 for any $H \subseteq K \subseteq \mathscr{G}$; then (7.6) holds for every $S = \{x, y\}$.

Let $K \subseteq H \subseteq \mathscr{G}$ and assume that $[H : K] = 2$. Let $\pi \in H$ be such that K, $\pi \cdot K$ are the two cosets of K in H. It follows from the definition of $\mathscr{G}$

that there is n_0 such that π is the identity on each $T_n, n \geqslant n_0$. Let $H^* = \{\rho \in H: \rho$ is the identity on each T_n, $n \geqslant n_0\}$, and $K^* = K \cap H^*$. Note that K^* and πK^* are the two cosets of K^* in H^* and so we have $[H^* : K^*] = [H : K] = 2$. However, $H^* \subseteq \{\rho \in \mathscr{G} : \rho$ is the identity on each T_n, $n \geqslant n_0\}$, and the latter group has order 3^{n_0}. The index $[H^* : K^*]$ should therefore be a divisor of 3^{n_0} which it is not. A contradiction.

To construct a model of ZF in which C_2 holds and C_3 fails, we use the First Embedding Theorem to get a symmetric model $\mathscr{N}$, and C_3 fails in the model because the statement '$\{T_n: n \in \omega\}$ has no choice function' will transfer from $\mathscr{V}$ to $\mathscr{N}$. We show that we can use Theorem 6.6 to make $\mathscr{N}$ a support model, and that the statement C_2 transfers to $\mathscr{N}$ by Theorem 6.7.

To be able to apply Theorem 6.6, we have to change the definition of a support. In $\mathscr{V}$, the supports will not be all finite subsets of A, but we restrict ourselves to sets of the form $E = T_{n_1} \cup \ldots \cup T_{n_k}$. With this restriction, one can easily see that if $\pi \in \text{fix}\,(E_1 \cap E_2)$, then $\pi = \rho \cdot \sigma$ for some $\rho \in \text{fix}\,(E_1)$, $\sigma \in \text{fix}\,(E_2)$. Now we can apply Theorem 6.6 to show that every $x \in \mathscr{N}$ has a least support (e, E), where e is a finite set of sets of ordinals and $E = \tilde{T}_{n_1} \cup \ldots \cup \tilde{T}_{n_k}$.

By Theorem 6.7, there is in $\mathscr{N}$ a one-to-one function F from $\mathscr{N}$ into $\mathscr{P}(\kappa) \times \mathscr{V}$, and $\mathscr{P}(\kappa)$ is linearly ordered. We define a function G on all finite sets in $\mathscr{N}$ as follows: If S is a finite set, $S = \{z_1, \ldots, z_n\}$, then let $(y_1, v_1) = F(z_1), \ldots, (y_n, v_n) = F(z_n)$ and let $e_S = \{y_1, \ldots, y_n\}$. The set e_S is finite and has a canonical enumeration $e_S = \{x_1, \ldots, x_k\}$ such that $x_1 < \ldots < x_k$. We let $G(S) = \{(m_1, v_1), \ldots, (m_n, v_n)\}$, where m_i is the image of y_i under the correspondence $x_1 \mapsto 1, \ldots, x_k \mapsto k$. It is obvious that the values of G are in $\mathscr{V}$ and that $G(S)$ has the same number of elements as S for every finite $S \in \mathscr{N}$.

Now it is clear that for any n, if C_n holds in $\mathscr{V}$ then C_n holds in $\mathscr{N}$, because G assigns to n-element sets in $\mathscr{N}$, n-element sets in $\mathscr{V}$. In particular, since C_2 holds in $\mathscr{V}$, we have C_2 in $\mathscr{N}$.

Example 2 illuminates the methods which are used in independence results concerning the Axiom of Choice for finite sets. We will now prove the first general result of this section.

PROOF OF THEOREM 7.11. We shall construct a permutation model $\mathscr{V}$ which satisfies C_n for each n but has a family of finite sets without a choice function. This will establish the independence result for the ZF set theory in view of the transfer method explained in Example 7.13, which works in this case as well.

Assume that the set of atoms A is countable and let $A = \bigcup_{n=0}^{\infty} P_n$, where $P_n = \{a_1^n, \ldots, a_{p_n}^n\}$, p_n being the nth prime number ($p_0 = 2$, $p_1 = 3$, $p_2 = 5$, etc.). Let $\mathscr{G}$ be the group generated by the following permutations π_n of A:

$$\text{(7.7)} \qquad \begin{aligned} &\pi_n\colon a_1^n \mapsto a_2^n \mapsto \ldots \mapsto a_{p_n}^n \mapsto a_1^n, \\ &\pi_n x = x \quad \text{for all } x \notin P_n. \end{aligned}$$

($\mathscr{G}$ is the weak direct product of cyclic groups of order p_n.) The model $\mathscr{V}$ is determined by $\mathscr{G}$ and the ideal of finite subsets of A.

We will follow the arguments in Example 7.13. To begin with, one can easily see that the family of $\{P_n : n \in \omega\}$ does not have a choice function. To show that for every n, every family of n-element sets in $\mathscr{V}$ does have a choice function, we shall prove an analogue of (7.6). For $n \in \omega$, let $\mathscr{G}_n = \operatorname{fix}(P_0 \cup \ldots \cup P_n)$; $\mathscr{G}_n$ is the weak direct product of cyclic groups of order p_k, $k > n$. It suffices to show that whenever $S \in \mathscr{V}$ has n elements, then

$$\text{(7.8)} \qquad \text{for some } x \in S,\ \operatorname{sym}(x) \supseteq \operatorname{sym}(S) \cap \mathscr{G}_n.$$

As in Example 7.13, if (7.8) is true and $\mathscr{F}$ is a family of n-element sets in $\mathscr{V}$, then we can construct a choice function f on $\mathscr{F}$ such that $\operatorname{sym}(f) \supseteq \operatorname{sym}(\mathscr{F}) \cap \mathscr{G}_n$.

To show that (7.8) holds, we will actually show that (7.8) holds for every $x \in S$. Let $x \in S$, let $H = \mathscr{G}_n \cap \operatorname{sym}(S)$ and $K = H \cap \operatorname{sym}(x)$. As in Example 7.13, if $K \neq H$, then the index $[H : K]$ is equal to the size of the orbit $\{\pi x : \pi \in H\}$. Since S has n elements, it is clear that it suffices to prove the following lemma:

LEMMA 7.14. *If $K \subseteq H \subseteq \mathscr{G}_n$, then either $K = H$ or $[H : K] > n$.*

PROOF. Assume that $[H : K] \leqslant n$. Then $H = K \cup \pi_1 \cdot K \cup \ldots \cup \pi_{n-1} \cdot K$, and there is k such that all $\pi_1, \ldots, \pi_{n-1}$ are the identity on each P_j, $j > k$. Similarly as in Example 7.13, we have $[H : K] = [H^* : K^*]$, where $K^* \subseteq H^* \subseteq \{\rho \in \mathscr{G}_n : \rho$ is the identity on each P_j, $j > k\}$ and the latter group has order $p_{n+1} \cdot \ldots \cdot p_k$. Thus $[H : K]$ would have to be a divisor of $p_{n+1} \cdot \ldots \cdot p_k$, which is impossible since we assumed $[H : K] \leqslant n < p_{n+1}$.

Theorems 7.15 and 7.16 deal with the relationship of the statements C_n, for different n.

Let m, n be two natural numbers. We say that m, n satisfy condition (S) if the following is true:

(S) *There is no decomposition of n in a sum of primes,*

$$n = p_1 + \dots + p_s,$$

such that $p_i > m$ *for all* $i = 1, \dots, s$.

(In particular, if $n > m$, then n is not a prime.)

THEOREM 7.15. *If* m, n *satisfy condition* (S) *and if* C_k *holds for every* $k \leqslant m$, *then* C_n *holds.*

PROOF. The proof follows closely the proof of $C_2 \to C_4$ in Example 7.12. We shall prove the theorem by induction on n. Let m be fixed, let m, n satisfy condition (S) and assume that the implication of the theorem is true for every $l < n$. Since n satisfies (S), n is not a prime and so n is divisible by some prime $p < n$. Necessarily, $p \leqslant m$, since otherwise we would have $n = p + \dots + p$, contrary to (S). We shall describe a way to choose from sets with n elements. Let A be an n-element set. Consider the set P of all p-element subsets of A; it has $\binom{n}{p}$ elements. A given choice function g chooses one $a \in A$ in each $X \in P$. For every $a \in A$, let $q(a)$ be the number of all $X \in P$ such that $g(X) = a$, and let q be the least such $q(a)$. We let $B = \{a \in A : q(a) = q\}$. Since $\binom{n}{p}$ is not divisible by n, it follows that both B and $A - B$ are nonempty. Let $|B| = l_1$ and $|A - B| = l_2$. We claim that either m, l_1 or m, l_2 satisfy condition (S). Otherwise, $l_1 = p_1 + \dots + p_r$ and $l_2 = p_{r+1} + \dots + p_s$, where $p_1, \dots, p_s$ are primes bigger than m, and then $n = p_1 + \dots + p_s$, contrary to the assumption. By the induction hypothesis, either C_{l_1} holds and we choose an element in B, or if C_{l_1} fails, then C_{l_2} holds and we choose an element in $A - B$.

THEOREM 7.16. *If* m, n *do not satisfy condition* (S), *then there is a model of set theory in which* C_k *holds for every* $k \leqslant m$ *but* C_n *fails.*

PROOF. Again, we shall construct a permutation model $\mathscr{V}$ and the result transfers to ZF, as explained in Example 7.13. Let m, n be two natural numbers which do not satisfy condition (S). That is,

$$n = p_1 + \dots + p_s,$$

where $p_1, \dots, p_s$ are primes, all bigger than m.

Assume that the set of atoms is countable and let $A = \bigcup_{i=0}^{\infty} N_i$, with $N_i = P_{i1} \cup \dots \cup P_{is}$ and $P_{ij} = \{a_{ij}^1, \dots, a_{ij}^{p_j}\}$. Let $\mathscr{G}$ be the group generated by the permutations

$$\pi_{ij}\colon a_{ij}^1 \mapsto a_{ij}^2 \mapsto \dots \mapsto a_{ij}^{p_j} \mapsto a_{ij}^1,$$
$$\pi_{ij} x = x \quad \text{for all } x \in A - P_{ij}.$$

The model $\mathscr{V}$ is determined by $\mathscr{G}$ and the ideal of finite subsets.

As in Theorem 7.11 or Example 7.13, one can easily see that the family $\{N_i: i \in \omega\}$ does not have a choice function, and so C_n fails in $\mathscr{V}$. To show that C_k holds in $\mathscr{V}$ for every $k \leqslant m$, it suffices to show that whenever $S \in \mathscr{V}$ has k elements, then

$$\text{sym}(x) \supseteq \text{sym}(S) \tag{7.9}$$

for each $x \in S$. Again, as in Theorem 7.11, it suffices to show that if $H \subseteq \mathscr{G}$ and $K \subseteq H$, then either $K = H$ or $[H:K] > m$. As in Lemma 7.14, one can prove that the index $[H:K]$ must be a divisor of some product of the primes $p_1, \ldots, p_s$, and consequently, $[H:K] > m$, since all the primes $p_1, \ldots, p_s$ exceed m.

7.5. Problems

1. For every partition of $[\omega]^2$ into two classes, there exists an infinite homogeneous set $H \subseteq \omega$.

[*Hint*: Let f: $[\omega]^2 \to \{0, 1\}$ be a partition. Let D be a nontrivial ultrafilter over ω. Define $f': \omega \to \{0, 1\}$ as follows: $f'(n) = 1$ if and only if $P_n = \{m > n: f(n, m) = 1\} \in D$; assume that $P = \{n: f'(n) = 1\} \in D$ (the proof is the same in the other case). Let $h_0 = \min(P)$, $h_1 = \min(P \cap P_{h_0})$, $h_2 = \min(P \cap P_{h_0} \cap P_{h_1})$, etc.; $H = \{h_0, h_1, \ldots\}$.]

2. Let r, k be given. For every partition of $[\omega]^r$ into k classes, there exists an infinite homogeneous set $H \subseteq \omega$.

[*Hint*: Let f: $[\omega]^r \to \{1, \ldots, k\}$ be a partition. Let D be nontrivial ultrafilter over ω. Define f': $[\omega]^{r-1} \to \{1, \ldots, k\}$ as follows: $f'(\{n_1, \ldots, n_{r-1}\}) = i$ if and only if $\{m > n_{r-1}: f(\{n_1, \ldots, n_{r-1}, m\}) = i\} \in D$, then define f'': $[\omega]^{r-2} \to \{1, \ldots, k\}$ similarly, etc.; finally, let $i \in \{1, \ldots, k\}$ be such that $\{n: f^{(r-1)}(n) = i\} \in D$. Every finite set $X \subset \omega$ such that $f(x) = i$ for all $x \in [X]^r$ can be extended to a larger X' with the same property and that gives an infinite homogeneous H.]

*3. Let r, k be given. For every z there is some w such that whenever $|W| \geqslant w$ and $[W]^r$ is partitioned into k classes, then there is a homogeneous set $H \subseteq W$ with at least z elements.

[*Hint*: Use the Compactness Theorem and Problem 2. Consider the theory with k r-ary predicates P_i $(i = 1, \ldots, k)$ and infinitely many constants c_n, $n = 0, 1, \ldots$, and consider the sentence saying that the P_i's constitute a partition of r-tuples and that no z elements form a homogeneous set, and the infinitely many sentences $c_n \neq c_m$ $(n \neq m)$. If the theorem is false, then any finite number of our sentences has a model, and by the Compactness Theorem, the set of all these sentences has a model. This defines a partition of $[\{c_n\}_{n=0}^{\infty}]^r$ which does not have a homogeneous set of size z; this contradicts Problem 2.]

*4. Let r_1, r_2, k be given. For every z, there is some w such that whenever $|W| \geqslant w$ and $[W]^{r_1} \times [W]^{r_2}$ is partitioned into k classes, then there are $Z_1, Z_2 \subseteq W$, $|Z_1| = |Z_2| = z$ such that $[Z_1]^{r_1} \times [Z_2]^{r_2}$ is included in one class.

[*Hint*: Use Problem 3. There is w_1 such that if $|W_1| \geqslant w_1$ and $[W_1]^{r_1}$ is partitioned into k classes, then there exists $Z_1 \subseteq W_1$, $|Z_1| = z$ such that $[Z_1]^{r_1}$ is included in one class. Let k_1 be the number of k-partitions of r_1-tuples in a w_1-element set. By Ramsey's Theorem, there is w_2 such that if $|W_2| \geqslant w_2$ and $[W_2]^{r_2}$ is partitioned into k_1 classes, then there exists $Z_2 \subseteq W_2$, $|Z_2| = z$ such that $[Z_2]^{r_2}$ is included in one class. Now take $w = \max(w_1, w_2)$.]

*5. Let $k, l, r_1, \ldots, r_l$ be given. For every z, there is some w such that whenever $|W| \geqslant w$ and $[W]^{r_1} \times \ldots \times [W]^{r_l}$ is partitioned into k classes, then there are $Z_1, \ldots, Z_l \subseteq W$, $|Z_1| = \ldots = |Z_l| = z$ such that $[Z_1]^{r_1} \times \ldots \times [Z_l]^{r_l}$ is included in one class.

[*Hint*: Use induction on l and Problem 4.]

**6. Let $\mathcal{N}$ be the model of ZF obtained by the First Embedding Theorem from the linearly ordered Mostowski model. Then the Prime Ideal Theorem holds in $\mathcal{N}$.

Notice that both this model $\mathcal{N}$ and the basic Cohen model establish the independence of the Axiom of Choice from the Prime Ideal Theorem. Notice also that the Selection Principle (which is a stronger version of the Ordering Principle) holds in the basic Cohen model while it fails in $\mathcal{N}$.

7. Let $(P, <)$ and $(Q, \prec)$ be countable universal homogeneous partially ordered sets. Then P and Q are isomorphic.

[*Hint*: Let $P = \{p_0, p_1, \ldots\}$, $Q = \{q_0, q_1, \ldots\}$. Define an isomorphism by a zig-zag argument, taking in turn p_n and q_n and finding an appropriate counterpart, using Lemma 7.5.]

*8. Let the forcing conditions $p \in P$ be finite 0, 1-functions defined on subsets of $\omega \times \omega$. As in the basic Cohen model, define canonical names $\underline{x}_1, \ldots, \underline{x}_n, \ldots$. Endow ω with a universal homogeneous partial order $<$ and let $\mathcal{G}$ be the group of all automorphisms of RO(P) induced by $<$-automorphisms of the set $\{\underset{\sim}{x}_1, ..., \underset{\sim}{x}_n, ...\}$; let $\mathscr{F}$ be generated by fix (e), e finite. Let $\mathscr{N}$ be the symmetric model given by $\mathscr{G}$ and $\mathscr{F}$. $\mathscr{N}$ satisfies the Ordering Principle, but the Prime Ideal Theorem fails in $\mathscr{N}$. Moreover, the Selection Principle holds in $\mathscr{N}$, and the partial ordering $<$ of the set $\{x_1, x_2, ..., x_n, ...\}$ cannot be extended to a linear ordering.

[*Hint*: As in Section 5.5, one proves that every $x \in \mathscr{N}$ has a least support, and that there is a one-to-one function F in $\mathscr{N}$ from the universe into $\text{On} \times I$, where I consists of finite sets of reals. This yields the Selection Principle. Assume that $<$ can be extended to a linear ordering $\prec$. Let $\prec$ be a name for $\prec$, with support e and let p_0 (with support e) force that $\prec$ is a linear ordering. Choose $\underset{\sim}{x}_1, ..., \underset{\sim}{x}_6$ outside e such that $\underset{\sim}{x}_4 < \underset{\sim}{x}_6$ and otherwise $<$-incomparable, and such that $p \supseteq p_0$ forces $\underset{\sim}{x}_1 \prec \underset{\sim}{x}_2 \prec \underset{\sim}{x}_3$. Let $p_{1,2} = p$: $(e \cup \{\underset{\sim}{x}_1, \underset{\sim}{x}_2\})$, $p_{2,3} = p$: $(e \cup \{\underset{\sim}{x}_2, \underset{\sim}{x}_3\})$. Let π, $\rho \in \text{fix}(e)$ be such that $\pi(\underset{\sim}{x}_1) = \underset{\sim}{x}_6$, $\pi(\underset{\sim}{x}_2) = \underset{\sim}{x}_5$ and $\rho(\underset{\sim}{x}_3) = \underset{\sim}{x}_4$. Then $\pi p_{1,2} \cup \rho p_{2,3}$ forces $\underset{\sim}{x}_6 \prec \underset{\sim}{x}_5 \prec \underset{\sim}{x}_4$, contrary to $\underset{\sim}{x}_4 < \underset{\sim}{x}_6$.]

Both this model $\mathscr{N}$ and the model from Section 7.2 establish the independence of the Prime Ideal Theorem from the Ordering Principle. Notice that the Selection Principle holds in $\mathscr{N}$ but fails in the other model.

9. The model from Problem 8 satisfies the Axiom of Choice for families of well-orderable sets.

[*Hint*: The same proof as in Problem 5.22.]

10. Let $\mathscr{N}$ be the model of ZF obtained by the First Embedding Theorem either from the linearly ordered Mostowski model, or from the model in Section 7.2. Then $\mathscr{N}$ satisfies the Axiom of Choice for families of well-orderable sets.

[*Hint*: The same proof as in Section 7.3; use the fact that each finite E has a fixed enumeration given by the linear ordering of A.]

11. If $\mathscr{V}$ is a permutation model of ZFA and if $\mathscr{V}$ satisfies the Axiom of Choice for families of finite sets, then $\mathscr{V}$ satisfies the Axiom of Choice for families of well-orderable sets.

[*Hint*: Let $\mathscr{V}$ be given by a group $\mathscr{G}$ and a filter $\mathscr{F}$. Let $\mathscr{S} \in \mathscr{V}$ be a family of well-orderable sets, let $A = \bigcup \{S: S \in \mathscr{S}\}$ and let $f \in \mathscr{V}$ be a choice function on finite subsets of $A \cup \mathscr{P}(A)$. Let $H = \text{sym}(f) \cap \text{sym}(\mathscr{S}) \in \mathscr{F}$. It suffices to show that if $S \in \mathscr{S}$ then there is $x \in S$ such that

$$\text{sym}(x) \supseteq H \cap \text{sym}(S); \tag{7.10}$$

for sufficiency, see (7.6) in Section 7.4. We show that (7.10) holds for each

$x \in S$. If not, then there is $\pi \in H \cap \operatorname{sym}(S)$ such that $\pi x \neq x$. Let $X = \{\pi^k x : k = \ldots, -3, -2, -1, 0, 1, 2, 3, \ldots\}$; X is a subset of S. If X is finite, then $\pi X = X$ and $\pi y \neq y$ for each $y \in X$, a contradiction since $\pi(f(X)) = (\pi f)(\pi X) = f(X)$. If X is infinite, let $X_1 = \{\pi^{2k} x : k = \ldots, -2, -1, 0, 1, 2, \ldots\}$ and $X_2 = \{\pi^{2k+1} x : k = \ldots, -2, -1, 0, 1, 2, \ldots\}$. Since S is well-orderable in $\mathscr{V}$, both X_1 and X_2 are in $\mathscr{V}$. We have $\pi X_1 = X_2$ and $\pi X_2 = X_1$, a contradiction since $\pi(f\{X_1, X_2\}) = (\pi f)(\pi\{X_1, X_2\}) = f(\{X_1, X_2\})$.]

**12. There is a model of ZF which satisfies the Ordering Principle (and hence the Axiom of Choice for families of finite sets) and which has a countable family of countable sets without a choice function.

**13. There is a model of ZF which satisfies the Axiom of Choice for families of countable sets but does not satisfy the Axiom of Choice for families of well-orderable sets.

Let S be a set and let T be a set of finite sequences $t = \langle s_0, s_1, \ldots, s_n \rangle$ of elements of S such that (a) $t' \subseteq t \in T$, implies $t' \in T$, (b) for each $n \in \omega$, T has an element of length n, and (c) if $t \in T$, then there are only finitely many (possibly none) elements of S such that $t^\frown s \in T$. The set T partially ordered by inclusion is called an *ω-tree.*

14. The Axiom of Choice for finite sets implies that every ω-tree has an infinite chain.

[*Hint*: Construct $t_0 \subseteq t_1 \subseteq \ldots \subseteq t_n \subseteq \ldots$ so that each t_n has infinitely many extensions in T.]

An *antichain* in a partially ordered set consists of mutually incomparable elements.

15. The model used in the proof of Theorem 7.11 has an ω-tree which has no infinite chain and no infinite antichain.

*16. If m and n satisfy condition (S), then $n < 8m^2$.

[*Hint*: The proof uses two facts from number theory. Let $n \geqslant 8m^2$. We find a decomposition of n into primes greater than m. By Bertrand's Postulate, there are primes p and q such that $m < p < 2m < q < 4m$. By another number theoretical fact, there are integers ξ, η such that $n = \xi \cdot p + \eta \cdot q$. If $\xi, \eta \geqslant 0$, then we have a decomposition $n = p + \ldots + p + q + \ldots + q$. Let $\eta < 0$. There is λ such that $0 \leqslant \eta + \lambda p < p$; let $\eta' = \eta + \lambda q$, $\xi' = \xi - \lambda q$. We have $n = \xi' \cdot p + \eta' \cdot q$ and $\eta' > 0$; show that $\xi' > 0$. This must be so because $\eta' \cdot q < pq < (2m)(4m) \leqslant n$.]

17. $C_2 \to C_n$ is provable if and only if $n = 1, 2$ or 4.
[*Hint*: Example 7.12 is the if part. For the only if part, use Theorem 7.16 and Problem 16, and verify the cases $n \leqslant 32 = 8 \cdot 2^2$.]

Let Z be a finite set of positive integers and let C_Z denote the conjunction of C_m, $m \in Z$. We say that Z, n satisfy condition (S) if for every decomposition of n into a sum of primes, $n = p_1 + \dots + p_s$, at least one p_i belongs to Z.

18. If Z, n satisfy condition (S), then C_Z implies C_n.
[*Hint*: The same proof as in Theorem 7.15.]

19. (C_3 and C_7) implies C_9.
[*Hint*: Verify condition (S).]

Let S_n be the group of all permutations on $\{1, \dots, n\}$. A subgroup $\mathscr{G}$ of S_n is *without fixed points* if each $i \leqslant n$ is moved by some $\pi \in \mathscr{G}$. We say that Z, n satisfy condition (D) if for every subgroup $\mathscr{G}$ of S_n without fixed points there is a subgroup H of $\mathscr{G}$ and a finite sequence $H_1, \dots, H_k$ of proper subgroups of H such that the sum of indices $[H : H_1] + \dots + [H : H_k]$ is in Z.

**20. Condition (D) is sufficient for the implication $C_Z \to C_n$.

**21. If Z, n do not satisfy condition (D), then there is a model of ZF in which C_Z holds and C_n fails.

22. It is not provable that (C_3, C_5 and C_{13}) implies C_{15}.
[*Hint*: Apply Condition (D).]

Consider the following *Cofinality Principle*: Every linearly ordered set has a cofinal well-ordered subset.

23. The Cofinality Principle together with the Order Extension Principle imply the Axiom of Choice.
[*Hint*: Show that every set can be well-ordered. Let S be a set; consider the family $\mathscr{F}$ of pairs (X, W), where $X \subseteq S$ is well-ordered by W, and the partial ordering of $\mathscr{F}$ given by $(X_1, W_1) \prec (X_2, W_2)$ if and only if $X_1 \subset X_2$. There is a linear ordering $<$ of $\mathscr{F}$ extending $\prec$ and a cofinal well-ordered sequence $p_0 < p_1 < \dots < p_\alpha < \dots$ $(\alpha < \lambda)$, of elements of $\mathscr{F}$. It is obvious that the set $Y = \bigcup \{X: (X, W) = p_\alpha \text{ for some } \alpha < \lambda\}$ can be well-ordered. Show that $Y = S$. If $a \in S - Y$, then $Y \cup \{a\}$ can be well-ordered by some W and hence for some α, $(Y \cup \{a\}, W) < p_\alpha = (X, U)$. This is a contradiction because $X \subset Y \cup \{a\}$ and $<$ extends $\prec$.]

**24. There is a model of ZF which satisfies the Cofinality Principle but does not satisfy the Order Extension Principle.

**25. The Prime Ideal Theorem is independent from the Hahn–Banach Theorem.

We shall outline the independence proof for ZFA:

*26. In the second Fraenkel model (of ZFA), the Prime Ideal Theorem fails while the Hahn–Banach Theorem holds.

[*Hint*: The Prime Ideal Theorem fails because the Axiom of Choice for pairs fails. For each n, let V_n be the class of all $x \in \mathscr{V}$ such that $P_0 \cup \ldots \cup P_n$ supports x. Let $E, p, \phi \in \mathscr{V}$ be, respectively, a real vector space, a sublinear functional on E and a linear functional on a subspace of E, $\phi(x) \leqslant p(x)$. Let n be such that $E, p, \phi \in V_n$. Using choice, extend the restriction $\phi | V_n$ to a $\psi \leqslant p$ on $E \cap V_n$; we have $\psi \in V_n$. Show that ψ extends to a χ on E such that $\chi \in V_n$, $\chi \leqslant p$. If $x \in V_{n+k}$ then $[x] = \{\pi x\colon \pi \in \text{fix}\,(P_0 \cup \ldots \cup P_n)\}$ has at most 2^k elements, $[x] = \{x_1, \ldots, x_{2^k}\}$, and $x_1 + \ldots + x_{2^k} \in V_n$. Let $\chi(x) = \psi(2^{-k}(x_1 + \ldots + x_{2^k}))$. Then χ extends ϕ because both χ, $\phi \in V_n$ and agree on $E \cap V_n$.]

**27. There is a model of ZF in which the Prime Ideal Theorem fails, but every partial ordering can be extended to a linear ordering.

7.6. Historical remarks

The independence of the Axiom of Choice from the Prime Ideal Theorem was proved by Halpern [1964] for ZFA (the present proof is somewhat different from Halpern's proof), and by Halpern and Levy (1967] for ZF, using a combinatorial result of Halpern and Läuchli [1966]. Ramsey's Theorem (Problems 1–3) is due to Ramsey [1929]; the generalization in Problems 4 and 5 is due to Rado [1954]. The result in Problem 6 is due to Pincus [1969, 1973a]. The independence of the Prime Ideal Theorem from the Ordering Principle was proved by Mathias [1967], using the model constructed in Problem 8; the observation in Problem 9 is also his. The model we used appears in Pincus [1969]. The results about universal homogeneous structures are due to Jónsson [1960]. The independence of the Ordering Principle from the Axiom of Choice for finite sets was proved by Läuchli [1964] for ZFA, and transferred to ZF by Pincus [1968]. The result of

Problem 11 is due to P. E. Howard.[1] The results of Problems 12 and 13 were announced by Pincus.

The independence result in Theorem 7.11 was proved by Levy [1962] for ZFA and transferred to ZF by Pincus [1969, 1973b], whose is also the observation in Problem 15.[2] Example 7.12 in Section 7.4 is due to Tarski. Conditions (S) and (D) were formulated by Mostowski [1945], who proved the independence result of Theorem 7.16 (for ZFA) and the sufficiency of Condition (D) (Problem 20); the result of Problem 16 is also his. The proof of the sufficiency of condition (S) given here is due to Szmielew [1947]. The independence result in Problem 21 was announced by Gauntt [1970].[3,4] For additional results concerning the Axiom of Choice for finite sets, see the papers of Wiśniewski and Zuckerman. The results of Problems 23 and 24 are due to Morris [1969].[5] The independence of the Prime Ideal Theorem from the Hahn–Banach Theorem was proved by Pincus [1972b]. The independence of the Prime Ideal Theorem from the Order Extension Principle (Problem 27) is due to Felgner [1971b].

[1] The result of Problem 11 contradicts the results announced by Levy [1963b]. Unfortunately, the construction presented there cannot be completed.

[2] The transfer to ZF was also claimed by Marek [1966] but the outlined method appears to be unsatisfactory and has not been published.

[3] A contradicting result was announced and later withdrawn by Truss [1970].

[4] The example in Problem 22 is a counterexample to another condition of Mostowski, who conjectured its sufficiency and singled out this example as a test case.

[5] The independence result contradicts the claim of Felgner [1969] that the Cofinality Principle implies the Axiom of Choice. An error has been found by Morris (see Felgner's corrections to [1969]).

CHAPTER 8

SOME WEAKER VERSIONS OF THE AXIOM OF CHOICE

8.1. The Principle of Dependent Choices and its generalization

The Principle of Dependent Choices was formulated in Section 2.4: If R is a relation on a nonempty set S such that for every $x \in S$ there exists $y \in S$ with xRy, then there is a sequence $\langle x_n : n = 0, 1, 2, \dots \rangle$ of elements of S such that

$$x_0 R x_1, \quad x_1 R x_2, \quad \dots, \quad x_n R x_{n+1}, \quad \dots .$$

We have also shown in Section 2.4 that the Principle of Dependent Choices implies the Countable Axiom of Choice and this in turn implies that every infinite set has a countable subset (in other words, every cardinal number is comparable with $\aleph_0$, i.e., $|X| \leqslant \aleph_0$ or $|X| \geqslant \aleph_0$ hold for any X). It will be shown in this chapter that neither of these implications can be reversed. In addition, we will generalize each of the three statements and investigate the relationship among them.

In the present section, we introduce the statements DC_κ, AC_κ and W_κ, where κ is an aleph, and prove some implications among these statements. In the next section, we present models showing that the implications proved in the present section are in a way best possible.

Let κ be an aleph. Consider the following statements:

DC_κ. *Let S be a nonempty set and let R be a binary relation such that for every $\alpha < \kappa$ and every α-sequence $s = \langle x_\xi : \xi < \alpha \rangle$ of elements of S there exists $y \in S$ such that sRy. Then there is a function $f: \kappa \to S$ such that for every $\alpha < \kappa$, $(f|\alpha) R f(\alpha)$.*

AC_κ. *Every family $\mathscr{X}$ of nonempty sets such that $|\mathscr{X}| = \kappa$ has a choice function.*

W_κ. *For every X, either $|X| \leqslant \kappa$ or $|X| \geqslant \kappa$.*

Obviously, $AC_{\aleph_0}$ is the Countable Axiom of Choice, and $W_{\aleph_0}$ says that every infinite set has a countable subset. Also, it is not difficult to see that $DC_{\aleph_0}$ is a reformulation of the Principle of Dependent Choices.

THEOREM 8.1.

(a). *If* $\kappa < \lambda$, *then* DC_λ *implies* DC_κ, AC_λ *implies* AC_κ, *and* W_λ *implies* W_κ.

(b). DC_κ *implies both* AC_κ *and* W_κ.

(c). $(\forall\kappa)DC_\kappa$ *is equivalent to* AC; $(\forall\kappa)W_\kappa$ *is equivalent to* AC.

(d). *If* κ *is singular, then* $(\forall\lambda < \kappa)DC_\lambda$ *implies* DC_κ, *and* $(\forall\lambda < \kappa)AC_\lambda$ *implies* AC_κ.

(e). *If* κ *is a limit cardinal and* $\kappa_0 = \text{cf}(\kappa)$, *then* AC_{κ_0} *and* $(\forall\lambda < \kappa)W_\lambda$ *implies* W_κ. *In particular,* $AC_{\aleph_0}$ *implies* $W_{\aleph_0}$.

PROOF. (a). Let $\kappa < \lambda$. It is clear that AC_λ implies AC_κ, and W_λ implies W_κ. Let S be a nonempty set and R a relation which satisfies the assumptions of DC_κ. One can extend R to apply to α-sequences for all $\alpha < \lambda$ (e.g., pick $y_0 \in S$ and let $s\bar{R}y_0$ for each α-sequence s if $\kappa \leqslant \alpha \leqslant \lambda$). Then use DC_λ to get $g: \lambda \to S$ and take $f = g|\kappa$.

(b). Assume DC_κ and let $\mathscr{X} = \{X_\alpha: \alpha < \kappa\}$ be a family of nonempty sets. We shall find a choice function on $\mathscr{X}$. Let $S = \bigcup_{\alpha<\kappa} X_\alpha$ and if $\alpha < \kappa$ and s is an α-sequence in S, then let sRy just in case $y \in X_\alpha$. The assumptions of DC_κ are satisfied and so we have $f: \kappa \to S$ such that $(f|\alpha)Rf(\alpha)$ for every $\alpha < \kappa$. Obviously, f determines a choice function on $\mathscr{X}$.

To show that DC_κ implies W_κ, let X be a set such that $|X| \nless \kappa$. We use DC_κ to show that $|X| \geqslant \kappa$. Let $S = X$, and if $\alpha < \kappa$ and s is an α-sequence in S, let sRy if and only if $y \notin \text{rng}(s)$. Since $|S| \nless \kappa$, such y always exists and so the assumptions of DC_κ are satisfied. The function f which we obtain from DC_κ is one-to-one, which proves that $|S| \geqslant \kappa$.

(c). If W_κ holds for every κ, then every set can be well-ordered and so AC holds. By (b), $(\forall\kappa)DC_\kappa$ implies $(\forall\kappa)W_\kappa$.

(d). Let κ be singular and let $\kappa_0 = \text{cf}(\kappa)$. Let $\kappa = \lim\{\alpha_\xi: \xi < \kappa_0\}$, where α_ξ is an increasing sequence of alephs. First we show that $(\forall\lambda < \kappa)DC_\lambda$ implies DC_κ. Let S be a nonempty set and let R be a relation such that for every $\alpha < \kappa$ and every α-sequence s in S, there exists $y \in S$ with sRy. We want to find $f: \kappa \to S$ such that $(f|\alpha)Rf(\alpha)$ for every $\alpha < \kappa$. Note that since we assume $(\forall\lambda < \kappa)DC_\lambda$, we can find, for any $\lambda < \kappa$, a function $g: \lambda \to S$ such that $(g|\alpha)Rg(\alpha)$. Let T be the set of all α_ξ-sequences s in S, for all $\xi < \kappa_0$. We shall define a relation $\bar{R}$ between γ-sequences in T, for $\gamma < \kappa_0$, and elements of T. If $\gamma < \kappa_0$ and t is a γ-sequence in T, then let s_t be the transfinite sequence in S obtained by concatenation of the terms of T;

s_t is an α-sequence in S for some $\alpha < \kappa$. If $\xi < \kappa_0$, t is a ξ-sequence in T, and if $z \in T$, then we let $t\bar{R}z$ if and only if z is an α_ξ-sequence in S and if $s_t{}^\frown(z|\eta)Rz(\eta)$ for all $\eta < \alpha_\xi$. By the remark we made about $(\forall\lambda < \kappa)\mathrm{DC}_\lambda$, there is always such $z \in T$, and so $\bar{R}$ satisfies the assumptions of DC_{κ_0}. Thus we have $g: \kappa_0 \to T$ with $(g|\xi)\bar{R}g(\xi)$ for all $\xi < \kappa_0$. It is now clear that if f is the concatenation of $g(\xi)$'s, f is a κ-sequence in S and satisfies $(f|\alpha)Rf(\alpha)$ for every $\alpha < \kappa$.

To show that $(\forall\lambda < \kappa)\mathrm{AC}_\lambda$ implies AC_κ, let $\mathscr{X} = \{X_\alpha : \alpha < \kappa\}$ be a family of nonempty sets; as before, let $\kappa = \lim\{\alpha_\xi : \xi < \kappa_0\}$. Let $\mathscr{S} = \{S_\xi : \xi < \kappa_0\}$, where each S_ξ is the set of all choice functions on $\{X_\alpha : \alpha < \alpha_\xi\}$. Since AC_λ holds for every $\lambda < \kappa$, each S_ξ is nonempty. By AC_{κ_0}, we choose one element in each S_ξ and then combine them to obtain a choice function on $\mathscr{X}$.

(e). Let κ be a limit cardinal, let $\kappa_0 = \mathrm{cf}(\kappa)$ and assume AC_{κ_0} and $(\forall\lambda < \kappa)\mathrm{W}_\lambda$. Let $\kappa = \lim\{\alpha_\xi : \xi < \kappa_0\}$, where α_ξ is an increasing sequence of alephs. Let X be a set such that $|X| \not< \kappa$. We want to show that $|X| \geqslant \kappa$. For every $\xi < \kappa_0$, let S_ξ be the set of all pairs (Y, W) such that $Y \subseteq X$, $|Y| = \alpha_\xi$ and W is a well-ordering of Y of type α_ξ. Since $|X| \not< \alpha_\xi$ and W_{α_ξ} holds, we have $|X| \geqslant \alpha_\xi$ and so S_ξ is nonempty. By AC_{κ_0}, we can choose one (Y, W) in each S_ξ. Then it is easy to see that the union of the Y's has cardinality κ and so $|X| \geqslant \kappa$.

THEOREM 8.2. *The statement* $(\forall\kappa)\mathrm{AC}_\kappa$ *implies the Principle of Dependent Choices.*

PROOF. Let S be a nonempty set and let R be a relation on S such that for every $x \in S$ there exists $y \in S$ such that xRy. We want to show that there is a sequence $\langle x_n : n \in \omega\rangle$ with x_nRx_{n+1} for all n. First note that there is no function f from S into ordinals with the property that $f(y) < f(x)$ whenever xRy. For, if f is such a function, let $x \in S$ be such that $f(x)$ has the least possible value; then there is $y \in S$ with xRy and so $f(y) < f(x)$.

We shall show that if AC_κ holds for every κ, then there is a sequence $\langle x_n : n \in \omega\rangle$ with x_nRx_{n+1} for all n, by contradiction. Assuming that there is no such sequence, we will construct a function f from S into ordinals such that $f(y) < f(x)$ whenever xRy. Let X be a well-orderable subset of S. Every subset Y of X contains an x such that xRy for no $y \in Y$; otherwise we could use a well-ordering of X to construct a sequence $\langle x_n : n \in \omega\rangle$ with x_nRx_{n+1} for all n. Thus the inverse of R is a well-founded relation on X and there is a unique ordinal α and a unique function $f_X : X \to \alpha$ such that $f_X(y) < f_X(x)$ if and only if xRy for all $x, y \in X$. Now we define f as follows:

$f(x) = \sup\{f_X(x): x \in X$ and X is a well-orderable subset of $S\}$. It is clear that f is a function from S into ordinals and that $f(y) \leqslant f(x)$ whenever xRy. To show that $f(y) < f(x)$ whenever xRy, it suffices to show that for every $x \in S$ there is a well-orderable set $X \subseteq S$ such that $f(x) = f_X(x)$.

Let $x \in S$ and let $f(x) = \alpha$. For every $\xi < \alpha$, let S_ξ be the set of all pairs (X, W) such that X is a well-orderable subset of S, W is a well-ordering of X, and $f_X(x) \geqslant \xi$. Each S_ξ is nonempty, and by $\mathrm{AC}_{|\alpha|}$, we can choose one (X, W) in each S_ξ. If Y is the union of the X's, then Y can easily be well-ordered, using the W's, and it is easy to see that $f_Y(x) \geqslant f_X(\xi)$ for all the X's. Thus $f_Y(x) = \alpha$.

8.2. Independence results concerning the Principle of Dependent Choices

In this section, we present four models which provide us with counter-examples to complete the study of interdependence started in Section 8.1. In Theorems 8.3, 8.6, 8.9, 8.12, we construct certain permutation models and outline how to transfer the results from ZFA to ZF. The first two cases are applications of the Second Embedding Theorem: The negative statements transfer automatically, and the positive statements, in this case DC_κ and W_κ, transfer from the permutation model into the symmetric model by a proper choice of a feasible normal filter. The transfer of Theorem 8.9 involves the statement $(\forall\kappa)\mathrm{AC}_\kappa$ and needs a further refinement of the Second Embedding Theorem.

We already mentioned that the implications in Theorems 8.1 and 8.2 are in a certain sense best possible. Let us be more specific:

(1). If κ is regular, then there is a model in which DC_λ (and hence also AC_λ and W_λ) holds for every $\lambda < \kappa$ but both AC_κ and W_κ (and also DC_κ) fail (Theorem 8.3). If κ is singular, then there is a model in which W_λ holds for every $\lambda < \kappa$ but W_κ fails (Theorem 8.6). This shows that (a) and (d) of Theorem 8.1 are best possible.

(2). For every κ there is a model in which both AC_κ and W_κ hold but $\mathrm{DC}_{\aleph_0}$ fails (Theorem 8.12). For every κ, there is a model in which W_κ holds and $\mathrm{AC}_{\aleph_0}$ fails (Theorem 8.6). There is a model in which $(\forall\kappa)\mathrm{AC}_\kappa$ holds and $\mathrm{W}_{\aleph_1}$ fails (Theorem 8.9). This shows that the implication in Theorem 8.1(b) cannot be reversed, that $\mathrm{AC}_{\aleph_0} \to \mathrm{W}_{\aleph_0}$ cannot be improved and that unlike DC_κ and W_κ in Theorem 8.1(c), $(\forall\kappa)\mathrm{AC}_\kappa$ does not imply the Axiom of Choice.

(3). Theorem 8.2 is also best possible because $(\forall\kappa)\mathrm{AC}_\kappa$ does not imply $\mathrm{DC}_{\aleph_1}$ (Theorem 8.9), and for every κ, there is a model in which AC_κ holds and $\mathrm{DC}_{\aleph_0}$ fails (Theorem 8.12).

THEOREM 8.3. *Let $\aleph_\alpha$ be a regular aleph. There exists a model in which:*

(i) DC_λ *holds for each* $\lambda < \aleph_\alpha$;

(ii) $AC_{\aleph_\alpha}$ *fails; there is a family of* κ *pairs without a choice function;*

(iii) $W_{\aleph_\alpha}$ *fails; there is a set A such that* $|A| > \lambda$ *for all* $\lambda < \aleph_\alpha$ *but* $|A| \not\geq \aleph_\alpha$.

PROOF. First we exhibit a permutation model $\mathscr{V}$ and then we outline the embedding of $\mathscr{V}$ in a symmetric model $\mathscr{N}$.

Assume that the set A of all atoms has cardinality $\aleph_\alpha$ and that A is a disjoint union of $\aleph_\alpha$ pairs $A = \bigcup_{\gamma<\aleph_\alpha} P_\gamma$, $P_\gamma = \{a_\gamma, b_\gamma\}$. Let $\mathscr{G}$ be the group of all permutations π of A with the property that for every $\gamma < \aleph_\alpha$, either $\pi a_\gamma = a_\gamma$ and $\pi b_\gamma = b_\gamma$, or $\pi a_\gamma = b_\gamma$ and $\pi b_\gamma = a_\gamma$. Let $\mathscr{F}$ be the filter on $\mathscr{G}$ generated by the groups

$$\text{fix}\,(E) = \{\pi \in \mathscr{G}: \pi x = x \text{ for all } x \in E\},$$

where $E \subseteq A$, $|E| < \aleph_\alpha$. The filter $\mathscr{F}$ is apparently normal and determines a permutation model $\mathscr{V}$. Every $x \in \mathscr{V}$ has a support $E \subseteq A$ such that $|E| < \aleph_\alpha$ and $\text{fix}\,(E) \subseteq \text{sym}\,(x)$.

To show that DC_λ holds in $\mathscr{V}$ for every $\lambda < \aleph_\alpha$, let $S \in \mathscr{V}$ and let $R \in \mathscr{V}$ be such that for every $\gamma < \lambda$ and every γ-sequence s in S there exists $x \in S$ such that sRx. We are looking for a function $f \in \mathscr{V}$ such that $(f|\gamma)Rf(\gamma)$ for all $\gamma < \lambda$.

LEMMA 8.4. *If* $\beta < \aleph_\alpha$ *and g is a function on* β *with values in* $\mathscr{V}$, *then* $g \in \mathscr{V}$.

PROOF. For every $\xi < \beta$, $g(\xi) \in \mathscr{V}$ has a support E_ξ such that $|E_\xi| < \aleph_\alpha$. Let $E = \bigcup_{\xi<\beta} E_\xi$. Clearly, $|E| < \aleph_\alpha$ and E is a support for each $g(\xi)$, $\xi < \beta$. Consequently, if $\pi \in \text{fix}\,(E)$, then $\pi g = g$ and so $\text{sym}\,(g) \supseteq \text{fix}\,(E)$, which means that $g \in \mathscr{V}$.

Now we can finish the proof of DC_λ in $\mathscr{V}$. By Lemma 8.4, for every $\gamma < \lambda$, each γ-sequence s in S belongs to $\mathscr{V}$ and so R satisfies (in the universe) the assumptions of DC_λ. Thus there exists $f: \lambda \to S$ such that $(f|\gamma)Rf(\gamma)$ for all $\gamma < \lambda$. By Lemma 8.4 again, f belongs to $\mathscr{V}$ and so DC_λ holds in $\mathscr{V}$.

It is easy to see that the enumeration $\langle P_\gamma: \gamma < \aleph_\alpha\rangle$ is in $\mathscr{V}$ and so the set of pairs $\{P_\gamma: \gamma < \aleph_\alpha\}$ has cardinality $\aleph_\alpha$ in $\mathscr{V}$. This set does not have a choice function in $\mathscr{V}$ since if f is a choice function, then there is no $E \subseteq A$, $|E| < \aleph_\alpha$, such that $\text{fix}\,(E) \subseteq \text{sym}\,(f)$: For every $|E| < \aleph_\alpha$, there exists $\gamma < \aleph_\alpha$ and $\pi \in \mathscr{G}$ such that $\pi \in \text{fix}\,(E)$ and $\pi a_\gamma = b_\gamma$ and hence $\pi f \neq f$. Thus $AC_{\aleph_\alpha}$ fails in $\mathscr{V}$.

Similarly, one can see that there is no one-to-one function f from $\aleph_\alpha$ into A: If $f: \aleph_\alpha \to A$ is one-to-one and $|E| < \aleph_\alpha$, then there exists $\pi \in \text{fix}\,(E)$

and $\gamma < \aleph_\alpha$ such that $\pi(f(\gamma)) \neq f(\gamma)$ and so $\pi f \neq f$. Hence $W_{\aleph_\alpha}$ fails in $\mathscr{V}$.

To get a symmetric model $\mathscr{N}$ of ZF in which Theorem 8.3 (i), (ii), (iii) hold, we use the Second Embedding Theorem and the feasible filter described in Problem 6.12.

Let us use the notation from Chapter 6. $\mathscr{U}$ is a model of ZFA+AC, $\mathscr{M}$ (the ground model) is the kernel of $\mathscr{U}$, $\mathscr{V} \subseteq \mathscr{U}$ is the permutation model described above. We choose some large regular cardinal κ and define $\mathscr{M}[G]$ as in the first Embedding Theorem, with $\tilde{a} = \{x_{a\xi}: \xi < \kappa\}$, as in (6.2). Then we have the embedding $\tilde{}$ of $\mathscr{U}$ in $\mathscr{M}[G]$. We define $\overline{\mathscr{G}}$ as in (6.6), and let $\overline{\mathscr{F}}$ be the filter on $\overline{\mathscr{G}}$ generated by

$$\{\overline{H}: H \in \mathscr{F}\} \cup \{\text{fix}\,(e): e \subseteq A \times \kappa \text{ and } |e| < \aleph_\alpha\}.$$

$\overline{\mathscr{F}}$ is a feasible filter and we may use the Second Embedding Theorem to show that $AC_{\aleph_\alpha}$ and $W_{\aleph_\alpha}$ fail in the corresponding symmetric model $\mathscr{N}$.

The proof of $(\forall\lambda < \aleph_\alpha)DC_\lambda$ in $\mathscr{N}$ is along the same lines as in $\mathscr{V}$ and is based on the following fact:

LEMMA 8.5. *If $\beta < \aleph_\alpha$ and $g \in \mathscr{M}[G]$ is a function on β with values in $\mathscr{N}$, then $g \in \mathscr{N}$.*

PROOF. First note that since $\beta < \kappa$ and so if $p_0 \subseteq p_1 \subseteq \dots p_\xi \subseteq \dots$, $\xi < \beta$, are forcing conditions, then $p = \bigcup_{\xi<\beta} p_\xi$ is a forcing condition, and so, as in Lemma 5.12, every β-sequence of elements of $\mathscr{M}$ which is in $\mathscr{M}[G]$ is in $\mathscr{M}$. A similar argument shows (see also Lemma 6.5) that there exists a function $f: \beta \to \mathscr{M}^B$, $f \in \mathscr{M}$, such that $g(\xi) = i_G(f(\xi))$ for each $\xi < \beta$. Even better, since g has a name $\underline{g} \in$ HS, the function f can be found such that $f: \beta \to$ HS. For each $\xi, f(\xi)$ has a support (E_ξ, e_ξ), where $E_\xi \subseteq A$ and $e_\xi \subseteq \bigcup_{a\in A} \tilde{a}$, and both $|E_\xi| < \aleph_\alpha$ and $|e_\xi| < \aleph_\alpha$. Let $E = \bigcup_{\xi<\beta} E_\xi$, $e = \bigcup_{\xi<\beta} e_\xi$ and let $\underline{f} \in \mathscr{M}^B$ be such that $[\![\underline{f}(\check{\xi}) = f(\xi)]\!] = 1$ for every $\xi < \beta$. Since $|E| < \aleph_\alpha$ and $|e| < \aleph_\alpha$, (E, e) is a support of $\underline{f}$ and so $\underline{f} \in$ HS. We have $g = i(\underline{f})$ and hence $g \in \mathscr{N}$.

Now, if $S \in \mathscr{N}$ and $R \in \mathscr{N}$ is a relation satisfying the assumptions of $DC_{\aleph_\alpha}$ in $\mathscr{N}$, then by Lemma 8.5, R satisfies the assumptions in $\mathscr{M}[G]$. Using the Axiom of Choice in $\mathscr{M}[G]$, we get a function $f: \aleph_\alpha \to S$, $f \in \mathscr{M}[G]$ such that $(f|\gamma)Rf(\gamma)$ for every $\gamma < \aleph_\alpha$. By Lemma 8.5, f belongs to $\mathscr{N}$. This completes the proof of Theorem 8.3.

THEOREM 8.6. *Let $\aleph_\alpha$ be a singular aleph. There exists a model in which:*

(i) DC_λ *holds for each* $\lambda < \text{cf}\,(\aleph_\alpha)$;

(ii) W_λ *holds for each* $\lambda < \aleph_\alpha$;

(iii) $W_{\aleph_\alpha}$ *fails and* $AC_{\text{cf}(\aleph_\alpha)}$ *fails.*

(Thus for every κ, there is a model in which W_κ holds and $AC_{\aleph_0}$ fails; it suffices to take $\aleph_\alpha > \kappa$ such that $\mathrm{cf}(\aleph_\alpha) = \aleph_0$.)

PROOF. Again, we exhibit first a permutation model $\mathscr{V}$ and then outline how the result transfers to ZF.

Let the set A of all atoms have cardinality $\aleph_\alpha$. Let $\mathscr{G}$ be the group of all permutations of A, and let $\mathscr{F}$ be the normal filter on $\mathscr{G}$ given by the ideal of sets of atoms of cardinality less than $\aleph_\alpha$. That is, $\mathscr{F}$ is generated by the groups $\mathrm{fix}(E)$, where $|E| < \aleph_\alpha$. Let $\mathscr{V}$ be the permutation model determined by $\mathscr{F}$.

It is easy to see that DC_λ holds in $\mathscr{V}$ for all $\lambda < \mathrm{cf}(\aleph_\alpha)$. This follows from the fact that if $\beta < \mathrm{cf}(\aleph_\alpha)$ and g is a function from β to $\mathscr{V}$, then $g \in \mathscr{V}$. The proof of this fact is the same as in Lemma 8.4. That DC_λ holds in $\mathscr{V}$ for all $\lambda < \mathrm{cf}(\aleph_\alpha)$ is deduced from that as in Theorem 8.3.

It is equally easy to see that $W_{\aleph_\alpha}$ fails in $\mathscr{V}$. The set A is not comparable by the cardinality to $\aleph_\alpha$. The proof is a routine use of permutations of A.

Once we show that W_λ holds for every $\lambda < \aleph_\alpha$, then by Theorem 8.1(e), we will have immediately that $AC_{\mathrm{cf}(\aleph_\alpha)}$ fails. This is because $W_{\aleph_\alpha}$ fails. Thus it remains to show that for every $\lambda < \aleph_\alpha$, W_λ holds in $\mathscr{V}$.

Let $\lambda < \aleph_\alpha$ and let $X \in \mathscr{V}$. We will show that $|X| \leqslant \lambda$ or $|X| \geqslant \lambda$. Let E_0 be a support of X, $E_0 \subseteq A$, $|E_0| < \aleph_\alpha$.

Case (i): For every $x \in X$, E_0 is a support of x. Then $\mathrm{fix}(X) \in \mathscr{F}$ and by (4.2), X can be well-ordered in $\mathscr{V}$.

Case (ii): Let $x \in X$ be such that E_0 is not a support of x. Let $E \subseteq A$ be such that $|E| < \aleph_\alpha$, $E \cap E_0 = 0$ and $E_0 \cup E$ is a support of x. Let $\{E_\xi : \xi < \lambda\}$ be a family of mutually disjoint subsets of A such that for each $\xi < \lambda$, $E_\xi \cap E_0 = 0$ and $|E_\xi| = |E|$. For each $\xi < \lambda$, let π_ξ be a permutation of A such that $\pi_\xi \in \mathrm{fix}(E_0)$ and $\pi_\xi'' E = E_\xi$; let $x_\xi = \pi_\xi(x)$. Let $Y = \{x_\xi : \xi < \lambda\}$. Since $\pi_\xi \in \mathrm{fix}(E_0) \subseteq \mathrm{sym}(X)$, we have $x_\xi \in X$ for each ξ, and so $Y \subseteq X$. Let $E' = E \cup \bigcup_{\xi<\lambda} E_\xi$. Clearly, $|E'| < \aleph_\alpha$ and E' is a support of Y so that $Y \in \mathscr{V}$. Actually, Y is well-orderable and the function $f: \lambda \to Y$ defined by $f(\xi) = x_\xi$ is in $\mathscr{V}$ because E' is a support of f. Thus if we prove that f is one-to-one, we will have $|X| \geqslant |Y| = \lambda$.

We show that $\pi_\xi x \neq \pi_\eta x$ whenever $\xi \neq \eta$. If $\pi_\xi x = \pi_\eta x$, then since $\pi_\xi''(E_0 \cup E) = E_0 \cup E_\xi$ and $\pi_\eta''(E_0 \cup E) = E_0 \cup E_\eta$, we conclude that both $E_0 \cup E_\xi$ and $E_0 \cup E_\eta$ are supports of x_ξ. It will follow from the following lemma that $E_0 = (E_0 \cup E_\xi) \cap (E_0 \cup E_\eta)$ is a support of E_ξ and so $E_0 = \pi_\xi^{-1} E_0$ is a support of $x = \pi_\xi^{-1} x_\xi$. This is a contradiction.

LEMMA 8.7. *Let E_1, E_2 be subsets of A, $|E_1| < \aleph_\alpha$, $|E_2| < \aleph_\alpha$. If $\pi \in \text{fix}(E_1 \cap E_2)$, then π is a product of permutations $\pi_1, \ldots, \pi_n$, each of them either in* fix (E_1) *or* fix (E_2). *Consequently, if $x \in \mathscr{V}$, and E_1 and E_2 are supports of x, then $E_1 \cap E_2$ is a support of x.*

PROOF. After the several lemmas of this kind, this should be a routine exercise for the reader ($n = 3$ is sufficient).

To transfer this result to ZF, we use the Second Embedding Theorem. We choose a regular cardinal κ, large enough to make sure that, say, $\mathscr{P}^\omega(A)^{\mathscr{V}} = \mathscr{P}^\omega(A)^{\mathscr{N}}$ under the embedding, and let $\overline{\mathscr{F}}$ be the feasible filter described in Problem 6.12. That is,

$$\tilde{a} = \{x_{a\xi} : \xi < \kappa\}$$

for each $a \in A$, and $\overline{\mathscr{F}}$ is the filter on $\overline{\mathscr{G}}$ generated by

$$\overline{\text{fix}(E)}, \quad E \subseteq A, \quad |E| < \aleph_\alpha$$

and

$$\text{fix}(e), \quad e \subseteq A \times \kappa, \quad |e| < \aleph_\alpha.$$

($\overline{\mathscr{G}}$ is defined as in (6.6).)

Let $\mathscr{N}$ be the symmetric model. We will show that Theorem 8.6 (i), (ii), (iii) hold in $\mathscr{N}$.

That $W_{\aleph_\alpha}$ fails in $\mathscr{N}$ is an immediate consequence of the Embedding Theorem.

To show that DC_λ holds for each $\lambda < \text{cf}(\aleph_\alpha)$, we notice that if $\beta < \text{cf}(\aleph_\alpha)$ and $g \in \mathscr{M}[G]$ is a function from β to $\mathscr{N}$, then $g \in \mathscr{N}$ (as in Lemma 8.5). Then the assertion follows.

Thus it remains to be shown that for every $\lambda < \aleph_\alpha$, W_λ holds in $\mathscr{N}$. Let $\lambda < \aleph_\alpha$ and let $X \in \mathscr{N}$. We will show that either $|X| \geqslant \lambda$ or there is $Y \in \mathscr{V}$ such that $|X| = |Y|$ in $\mathscr{N}$. Since W_λ holds in $\mathscr{V}$, we will be done.

Let $\underset{\sim}{\Delta}$ consist of all Boolean pairs $((\underset{\sim}{e}, \underset{\sim}{E})^B, \underset{\sim}{x})^B$ such that $(\underset{\sim}{e}, \underset{\sim}{E})$ is a support of $\underset{\sim}{x}$; i.e., $\text{fix}(\underset{\sim}{e}) \cap \overline{\text{fix}(\underset{\sim}{E})} \subseteq \text{sym}(\underset{\sim}{x})$. $\underset{\sim}{\Delta}$ is a symmetric class and so its interpretation Δ is a class of $\mathscr{N}$. For $e \subseteq \bigcup A = \{x_{a\xi} : a \in A, \xi < \kappa\}$, $|e| < \aleph_\alpha$ and $E \subseteq A$, $|E| < \aleph_\alpha$, we denote by $\Delta(e, E)$ the class (in $\mathscr{N}$) of all $x \in \mathscr{N}$ such that $((e, E), x) \in \Delta$. Let $\underset{\sim}{X}$ be a name for X and let $(\underset{\sim}{e}_0, \underset{\sim}{E}_0)$ be a support of $\underset{\sim}{X}$. We distinguish two cases:

Case (i): For every $x \in X$ there exists E such that $x \in \Delta(e_0, E)$. As usual, there is a function e in $\mathscr{N}$ such that for each $x \in \mathscr{N}$,

$$x = e(t, \underset{\sim}{y}),$$

where t is an assignment and $\underset{\sim}{y}$ is a symmetric name. In this context, an *assignment* is a pair $t = (t_1, t_2)$ such that t_1 is a one-to-one function on some e and t_2 is a one-to-one function on $\underset{\sim}{E}$, and for some $\pi \in \overline{\mathscr{G}}$, $t_1(\underset{\sim}{x}) = i(\pi \underset{\sim}{x})$ and $t_2(\underset{\sim}{a}) = i(\pi \underset{\sim}{a})$.

Let us fix some t_1 such that $e_0 = \operatorname{rng}(t_1)$ and for every $x \in X$, consider the pair $(\underset{\sim}{y}, w(x))$, where $\underset{\sim}{y}$ is the least element of HS such that there exists t_2 with $x = e((t_1, t_2), \underset{\sim}{y})$ and $w(x)$ is the set of all t_2 such that $x = e((t_1, t_2), y)$; let Y be the set of all these pairs. Notice that the correspondence between X and Y is one-to-one, so that $|X| = |Y|$ in $\mathscr{N}$ and $Y \in \mathscr{U}$. Actually, each t_2 represents a well-ordering of a subset of A; and since $\mathscr{P}^{\omega}(A)^{\mathscr{V}} = \mathscr{P}^{\omega}(A)^{\mathscr{N}}$, it can be concluded that $Y \in \mathscr{V}$.

Case (ii): There is some $x \in X$ such that $x \notin \Delta(e_0, E)$ for any E. We will show that X has a subset of cardinality λ. The method we use is similar to the proof of Theorem 6.6. Let $\underset{\sim}{x}$ be a name for x and let $\underset{\sim}{x} \in \underset{\sim}{\Delta}(\underset{\sim}{e}, \underset{\sim}{E})$ for some $e \supseteq e_0$. As in (6.19), we have:

$$\text{if } \underset{\sim}{y} \in \underset{\sim}{\Delta}(\underset{\sim}{e}, \underset{\sim}{E}) \text{ and } p \Vdash \phi(\underset{\sim}{y}), \text{ then } p{:}\,\underset{\sim}{e} \Vdash \phi(\underset{\sim}{y}),$$

and so it suffices to find $Y = \{y_\xi : \xi < \lambda\} \in \mathscr{N}$ such that: If $p \Vdash \underset{\sim}{x} \notin \underset{\sim}{\Delta}(\underset{\sim}{e}_0, \underset{\sim}{E})$, and $p = p : \underset{\sim}{e}$, then there is $q \supseteq p$ such that $q \Vdash \underset{\sim}{y}_\xi \in \underset{\sim}{X}$ for every $\xi < \lambda$, and $q \Vdash \underset{\sim}{y}_\xi \neq \underset{\sim}{y}_\eta$ whenever $\xi \neq \eta$. There exist permutations $\pi_\xi \in \overline{\mathscr{G}}$, $\xi < \lambda$ such that π_0 is the identity, $\pi_\xi(a) = a$ for each $a \in A$, $\pi_\xi \in \operatorname{fix}(\underset{\sim}{e}_0)$ and $\pi_\xi \underset{\sim}{e} \cap \pi_\eta \underset{\sim}{e} = \underset{\sim}{e}_0$ whenever $\xi \neq \eta$; moreover, $\underset{\sim}{e}^* = \bigcup \{\pi_\xi \underset{\sim}{e} : \xi < \lambda\}$ has cardinality $< \aleph_\alpha$. Let $\underset{\sim}{y}_\xi = \pi_\xi \underset{\sim}{x}$ and $Y = \{y_\xi : \xi < \lambda\}$. We have $\underset{\sim}{Y} \in \underset{\sim}{\Delta}(e \cup \underset{\sim}{e}^*, \underset{\sim}{E})$. If p is as above, let $q = \bigcup \{\pi_\xi p : \xi < \lambda\}$; q is a condition, $q \supseteq p$, and $q \Vdash \pi_\xi \underset{\sim}{x} \in \underset{\sim}{X}$ for every $\xi < \lambda$ because $\pi_\xi \in \operatorname{fix}(\underset{\sim}{e}_0) \cap \overline{\operatorname{fix}(\underset{\sim}{E}_0)}$, so that $\pi_\xi \underset{\sim}{X} = \underset{\sim}{X}$. We will be done if we can prove that $q \Vdash \pi_\xi \underset{\sim}{x} \neq \pi_\eta \underset{\sim}{x}$ whenever $\xi \neq \eta$. For this, we use a lemma whose proof follows closely either Theorem 6.6 or Lemma 5.23.

LEMMA 8.8 *If $\underset{\sim}{x}_1$ has a support $(\underset{\sim}{e}_1, \underset{\sim}{E}_1)$ and $\underset{\sim}{x}_2$ has a support $(\underset{\sim}{e}_2, \underset{\sim}{E}_2)$ and if $q \Vdash \underset{\sim}{x}_1 = \underset{\sim}{x}_2$, then there is $\underset{\sim}{z}$ with support $(\underset{\sim}{e}_1 \cap \underset{\sim}{e}_2, \underset{\sim}{E}_1 \cap \underset{\sim}{E}_2)$ such that $q \Vdash \underset{\sim}{x}_1 = \underset{\sim}{z}$.*

Now, if q_1 is a stronger condition than q and forces $\pi_\xi \underset{\sim}{x} = \pi_\eta \underset{\sim}{x}$ for some $\xi \neq \eta$, then $q_1 \Vdash \pi_\xi \underset{\sim}{x} \in \underset{\sim}{\Delta}(\underset{\sim}{e}_0, \underset{\sim}{E})$ and $\pi_\xi^{-1} q_1 \Vdash \underset{\sim}{x} \in \underset{\sim}{\Delta}(\underset{\sim}{e}_0, \underset{\sim}{E})$. However, $\pi_\eta^{-1} q_1 \supseteq p$, a contradiction.

THEOREM 8.9. *Let $\aleph_{\alpha+1}$ be a successor aleph. There exists a model in which:*

(i) $DC_{\aleph_\alpha}$ *holds*;
(ii) *both* $DC_{\aleph_{\alpha+1}}$ *and* $W_{\aleph_{\alpha+1}}$ *fail*;
(iii) AC_λ *holds for every* λ.

(*In particular,* $(\forall\lambda)AC_\lambda$ *does not imply* $DC_{\aleph_1}$ *or* $W_{\aleph_1}$.)

PROOF. First we exhibit a permutation model $\mathscr{V}$ and then outline the transfer into ZF.

Let the set A of atoms have cardinality $\aleph_{\alpha+1}$. Let $\mathscr{G}$ be the group of all permutations of A and let $\mathscr{F}$ be the normal filter on $\mathscr{G}$ given by the ideal of sets of atoms of cardinality at most $\aleph_\alpha$. Let $\mathscr{V}$ be the permutation model determined by $\mathscr{F}$. Each $x \in \mathscr{V}$ has a support $E \subseteq A$ such that $|E| \leqslant \aleph_\alpha$.

It is easy to check that functions from ω_α to $\mathscr{V}$ belong to $\mathscr{V}$ and so $DC_{\aleph_\alpha}$ holds in $\mathscr{V}$. It is even easier to see that the set A has no subset of cardinality $\aleph_{\alpha+1}$, and so $W_{\aleph_{\alpha+1}}$ fails in $\mathscr{V}$; consequently, $DC_{\aleph_{\alpha+1}}$ fails in $\mathscr{V}$.

To show that AC_λ holds in $\mathscr{V}$ for each λ, let $\mathscr{X} = \{X_\gamma : \gamma < \lambda\}$ be a family of nonempty sets, $\mathscr{X} \in \mathscr{V}$. We find a choice function $f \in \mathscr{V}$ on $\mathscr{X}$. Let E_0 be a support of $\mathscr{X}$ and let E_1 be a subset of A such that $E_1 \cap E_0 = 0$ and $|E_1| = \aleph_\alpha$. We will show that for each $\gamma < \lambda$ there is $x \in X_\gamma$ such that $x \in \Delta(E_0 \cup E_1)$. If we choose one such x for each γ, then the choice function f will have support $E_0 \cup E_1$ (because $X_\gamma \in \Delta(E_0)$ for each γ) and so $f \in \mathscr{V}$. Let $\gamma < \lambda$, and let y be an arbitrary element of X_γ. There is some E such that $y \in \Delta(E)$. There exists a permutation $\pi \in \mathscr{G}$ such that $\pi \in \operatorname{fix}(E_0)$ and $\pi''E \subseteq E_0 \cup E_1$. Hence $\pi y \in \pi X_\gamma = X_\gamma$, and since $E_0 \cup E_1 \supseteq \pi''E$ is a support of πy, we can take $x = \pi y$.

Now we briefly sketch the transfer of the above results into ZF. There is no problem with the transfer of $\neg W_{\aleph_{\alpha+1}}$ and $DC_{\aleph_\alpha}$. It seems, however, that the Second Embedding Theorem is too weak to enable to transfer $(\forall\lambda)C_\lambda$. We introduce a further refinement of the Embedding Theorem.

In the Second Embedding Theorem, one chooses a large regular cardinal $\kappa > |A|$, and lets $\tilde{a} = \{x_{a\xi} : \xi < \kappa\}$ for each $a \in A$, where $x_{a\xi}$ are subsets of κ (obtained from a generic subset of $A \times \kappa$). In the present proof, we will use a slightly different embedding. If $s \subseteq \kappa$, $|s| \leqslant \aleph_\alpha$, consider for each $a \in A$ and $\xi < \kappa$ the following subset $x_{a\xi s}$ of κ

$$\eta \in x_{a\xi s} \leftrightarrow \text{either } \eta \in x_{a\xi} \text{ and } \eta \notin s \\ \text{or } \eta \notin x_{a\xi} \text{ and } \eta \in s.$$

The canonical name for each $x_{a\xi s}$ is of course obtained from the canonical name for $x_{a\xi}$ by interchanging 0's and 1's at every $\eta \in s$. Now we let for each $a \in A$,

$$\tilde{a} = \{x_{a\xi s} : \xi < \kappa \text{ and } s \subseteq \kappa, |s| \leqslant \aleph_\alpha\}. \tag{8.1}$$

We shall define the group $\overline{\mathscr{G}}$ a little differently from (6.6). Let X be a subset of $(A \times \kappa) \times \kappa$, $|X| \leqslant \aleph_\alpha$. By σ_X we denote the automorphism of B which interchanges 0's and 1's at every point of X. More exactly, if p is a condition, let $(\sigma_X p)(a, \xi, \eta) = 1 - p(a, \xi, \eta)$ whenever $(a, \xi, \eta) \in X$ and $(\sigma_X p)(a, \xi, \eta) = p(a, \xi, \eta)$ otherwise. (Compare with Problem 5.24.) Now for each $\rho \in \mathscr{G}$, we let (as before) $[\rho]$ be the group of all permutations π of $A \times \kappa$ such that $\pi(a, \xi) = (\rho(a), \xi')$ for some ξ' (cf. (6.5)), and we let $\overline{\mathscr{G}}$ be generated by $\bigcup \{[\rho]: \rho \in \mathscr{G}\}$ and $\{\sigma_X: X \subseteq (A \times \kappa) \times \kappa, |X| \leqslant \aleph_\alpha\}$. Then we let $\overline{\mathscr{F}}$ be the filter on $\overline{\mathscr{G}}$ generated by $\overline{\text{fix}\,(E)}$, $E \subseteq A$, $|E| \leqslant \aleph_\alpha$, and fix (e), $e \subseteq A \times \kappa$, $|e| \leqslant \aleph_\alpha$, where

(8.2) $\text{fix}\,(e) = \{\pi \in \overline{\mathscr{G}}: \pi(\underline{x}_{a\xi s}) = \underline{x}_{a\xi s}$ for every $(a, \xi) \in e$ and every $s\}$.

(Note that (8.2) means that $\sigma_X \in \text{fix}\,(e)$ if and only if $X \cap (e \times \kappa) = 0$.) The filter $\overline{\mathscr{F}}$ is a normal filter on $\overline{\mathscr{G}}$ and $\tilde{A}$ has a symmetric name. If one goes through the proof of the First Embedding Theorem, one can verify that the same result holds in the present context. In particular, one can choose $\kappa > \aleph_\alpha$ so large as to get $\mathscr{P}^\omega(A)^{\mathscr{V}} = \mathscr{P}^\omega(A)^{\mathscr{N}}$. The failure of $W_{\aleph_{\alpha+1}}$ transfers from $\mathscr{V}$ to $\mathscr{N}$ easily, and so does $DC_{\aleph_\alpha}$, due to the choice of $\overline{\mathscr{F}}$ (if $g: \omega_\alpha \to \mathscr{N}$, $g \in \mathscr{M}[G]$, then $g \in \mathscr{N}$).

It remains to show that AC_λ holds in $\mathscr{N}$ for every λ. Let λ be fixed and let $\mathscr{X} = \{X_\gamma: \gamma < \lambda\}$ be a family of nonempty sets in $\mathscr{N}$. We are looking for a choice function $f \in \mathscr{N}$ on $\mathscr{X}$. Let $\underline{\Delta}$ be the support relation, with $\underline{E} \subseteq A$ and $\underline{e} \subseteq A \times \kappa$ (convention: $e = \{x_{a\xi s}: (a, \xi) \in \underline{e}, s \subseteq \kappa, |s| \leqslant \aleph_\alpha\}$).

LEMMA 8.10. *There exists $E_0 \subset \tilde{A}$, $|E_0| \leqslant \aleph_\alpha$ with the following property: For every $\gamma < \lambda$ there exists an $x \in X_\gamma$ such that for some e, $x \in \Delta(e, E_0)$.*

PROOF. For each γ, let Y_γ be the set of all pairs (E, W) such that $E \subseteq \tilde{A}$, W is a well-ordering of E and there exists an $x \in X_\gamma$ such that for some e, $x \in \Delta(e, E)$. The sets Y_γ are nonempty for each $\gamma < \lambda$ and since $Y = \{Y_\gamma: \gamma < \lambda\} \in \mathscr{P}^\omega(\tilde{A})^{\mathscr{N}}$, it follows that $Y \in \mathscr{V}$. Thus there is a choice function on Y. Take the union of all the chosen E's; this set E_0 can be well-ordered (using the chosen W's) and since it is a subset of $\tilde{A}$, it follows that $E_0 \in \mathscr{V}$ and $|E_0| \leqslant \aleph_\alpha$. Obviously, this E_0 will do.

Let $\underline{\mathscr{X}}$ be a name for $\mathscr{X}$ and let $(\underline{e}_0, \underline{E}_0)$ be a support of $\mathscr{X}$; we may assume that E_0 satisfies the assertion of Lemma 8.10. Let $\underline{e}_1 \subseteq A \times \kappa$ be such that $\underline{e}_1 \cap \underline{e}_0 = 0$ and that for each $a \in \underline{E}_0$, $|\underline{e}_1 \cap (\{a\} \times \kappa)| = \aleph_\alpha$.

LEMMA 8.11. *For every $\gamma < \lambda$, there exists an $x \in X_\gamma$ such that $x \in \Delta(e_0 \cup e_1, E_0)$.*

PROOF. Let $\underline{y} \in \underline{\Delta}(\underline{e}, \underline{E}_0)$ and let $p \Vdash \underline{y} \in \underline{X}_\gamma$. We may assume that $p = p : \underline{e}$. We shall find a stronger condition q which forces $\underline{y} \in \underline{\Delta}(\underline{e}_0 \cup \underline{e}_1, \underline{E}_0)$. There is $\pi \in \overline{\mathscr{G}}$ such that $\pi \in \text{fix}\,(\underline{e}_0)$, $\pi(a) = a$ for all $a \in A$ and $\pi'\underline{e} \subseteq \underline{e}_0 \cup \underline{e}_1$.

However, p and πp may not be compatible. Let X be the set of all places where p and πp disagree; let $\sigma = \sigma_X$. Then p and $\sigma\pi p$ are compatible. Since $X \cap (\underline{e}_0 \times \kappa) = 0$, we have $\sigma \in \operatorname{fix}(e_0)$ and so $\sigma\underline{\mathscr{X}} = \underline{\mathscr{X}}$. Also, we have $\sigma\pi\check{y} = \check{y}$, and so

$$\sigma\pi p \Vdash \sigma\pi\underline{y} \in \underline{X}_\gamma.$$

Since $\sigma\pi\underline{y} \in \underline{\Delta}(\underline{e}_0 \cup \underline{e}_1, \underline{E}_0)$, we have

$$p \cup \sigma\pi p \Vdash (\exists x \in \underline{X}_\gamma) x \in \underline{\Delta}(\underline{e}_0 \cup \underline{e}_1, \underline{E}_0).$$

Now we fix an assignment t with range $e_0 \cup e_1$ and E_0 and define (in $\mathcal{N}$) a choice function f on $\mathscr{X}$ as follows: $f(\gamma) = x$, where $x = e(t, \underline{y})$ with $\underline{y}$ being the least such name (such $\underline{y}$ exists by Lemma 8.11).

THEOREM 8.12. *Let κ be a regular cardinal. There exists a model in which:*
  (i) *the Principle of Dependent Choices fails;*
  (ii) AC_λ *holds for each* $\lambda < \kappa$;
  (iii) W_λ *holds for each* $\lambda < \kappa$.
(*In particular,* $\mathrm{AC}_{\aleph_0}$ *does not imply* $\mathrm{DC}_{\aleph_0}$.)

PROOF. Assume that $|A| = \kappa$ and identify the elements of A with finite sequences $s = \langle \xi_0, \xi_1, \ldots, \xi_{n-1}\rangle$ of ordinals smaller than κ. The set A is partially ordered by the extension of sequences, and we let $\mathscr{G}$ be the set of all automorphisms of $(A, \subseteq)$; that is, $\pi(s)$ has the same length as s, and $s \subseteq t$ implies $\pi(s) \subseteq \pi(t)$. Let I be the following ideal of subsets of A: A set $E \subseteq A$ is in I if:
  (i) $|E| < \kappa$;
  (ii) $s \subseteq t$ and $t \in E$ implies $s \in E$;
  (iii) E has no infinite branch; i.e., there is no $f: \omega \to \kappa$ such that $f|n \in E$ for all n.

Let $\mathscr{F}$ be the filter on $\mathscr{G}$ generated by $\{\operatorname{fix}(E): E \in I\}$. Let $\mathscr{V}$ be the permutation model given by $\mathscr{F}$.

It is easy to see that the principle of dependent choices fails in $\mathscr{V}$. The relation $s \subset t$ on A satisfies the assumptions of the principle but there is no infinite sequence $\{s_n: n \in \omega\} \in \mathscr{V}$ of elements of A such that $s_0 \subset s_1 \subset \ldots \subset s_n \subset \ldots$. This is because for every $E \in I$ there exists $\pi \in \operatorname{fix}(E)$ such that $\pi s_n \neq s_n$ for some n.

To show that AC_λ holds in $\mathscr{V}$ for every $\lambda < \kappa$, let $\mathscr{X} = \{X_\gamma: \gamma < \lambda\} \in \mathscr{V}$ be a family of nonempty sets, for some $\lambda < \kappa$. We want to find a choice function $f \in \mathscr{V}$ on $\mathscr{X}$. Let $E \in I$ be a support of $\mathscr{X}$. Notice that $\operatorname{sym}(X_\gamma) \supseteq \operatorname{fix}(E)$ for each $\gamma < \lambda$. For each $\gamma < \lambda$, choose some $y_\gamma \in X_\gamma$

and some $E_\gamma \in I$ such that $E_\gamma \supseteq E$ and $y_\gamma \in \Delta(E_\gamma)$. By recursion on γ, one can find automorphisms $\pi_\gamma \in \mathscr{G}$ such that $\pi_\gamma \in \text{fix}\,(E)$ for each γ, and $\pi_\gamma(E_\gamma) \cap \pi_\delta(E_\delta) = E$ whenever $\gamma \neq \delta$ (exercise). Let $x_\gamma = \pi_\gamma(y_\gamma)$ for each $\gamma < \lambda$. Let

$$E' = \bigcup \{\pi_\gamma(E_\gamma): \gamma < \lambda\}.$$

The reader can easily verify that $E' \in I$. It follows that E' is a support of each x_γ, and that $x_\gamma = \pi_\gamma y_\gamma \in \pi_\gamma X_\gamma = X_\gamma$. Hence the function f defined on $\mathscr{X}$ by $f(X_\gamma) = x_\gamma$ is in $\Delta(E')$ and so belongs to $\mathscr{V}$.

To show that W_λ holds in $\mathscr{V}$ for every $\lambda < \kappa$, we proceed as in the earlier theorems of this section. Let $\lambda < \kappa$, and let $X \in \mathscr{V}$ with a support E_0. Either E_0 is a support for every $x \in X$ and then X can be well-ordered, or we take $x \in X$ with a support $E \supseteq E_0$ such that $x \notin \Delta(E_0)$ and find λ automorphisms π_ξ, $\xi < \lambda$, such that $\pi_\xi \in \text{fix}\,(E_0)$ and $\pi_\xi(E) \cap \pi_\eta(E) = E_0$ whenever $\xi \neq \eta$. Then $\pi_\xi x \in X$ for each ξ and the function $f\colon \xi \mapsto \pi_\xi x$ has support $\bigcup \{\pi_\xi E: \xi < \lambda\}$. To show that $\pi_\xi x \neq \pi_\eta x$ whenever $\xi \neq \eta$, one uses the, by now notorious, lemma which says that if E_1 and E_2 are both supports of some z, then $E_1 \cap E_2$ is also a support of z. (Compare with the proof of W_λ in Theorem 8.6.)

To transfer this result to ZF, one can use the methods outlined in the proof of Theorems 8.3, 8.6, 8.9.

8.3. Problems

1. There is a model in which AC_λ holds for every λ, but for finite n there is a family of n-element sets without a choice function.

[*Hint*: Show that in the model in Theorem 8.9 the family of all n-element subsets of A has no choice function.]

*2. Assume $2^{\aleph_0} = \aleph_1$. There is a model in which (i) AC_λ holds for all λ, (ii) $W_{\aleph_1}$ holds, (iii) $W_{\aleph_2}$ fails, and (iv) $DC_{\aleph_1}$ fails.

[*Hint*: Let $|A| = \aleph_1$, and order A as the real numbers. Let $\mathscr{G}$ be the group of all order-preserving permutations, and let $\mathscr{F}$ be given by the ideal of intervals $(-\infty, a)$. Let $\mathscr{V}$ be the permutation model. Proofs in $\mathscr{V}$: (i) The same idea as in Theorem 8.9. (ii) The same idea as in Theorem 8.6. (iii) $|A| > \aleph_1$ in $\mathscr{V}$ but not $\leqslant \aleph_2$. (iv) Let R be the following relation on A: sRx if and only if $x > \sup\,(s)$. If $s \in \mathscr{V}$, then s is bounded, but there is no increasing ω_1-sequence of elements of A (in $\mathscr{V}$ or otherwise) and so R is a counterexample.]

3. $W_{\aleph_0}$ implies that every countable family of finite sets has a choice function.

*4. There is a model which satisfies $W_{\aleph_0}$ and which has a countable family of countable sets without a choice function.

[*Hint*: Let $\{A_i : i \in \omega\}$ be a disjoint family of countable sets and let $A = \bigcup_{i=0}^{\infty} A_i$. Let $\mathscr{G}$ be the set of all permutations of A which preserve each A_i; let $\mathscr{F}$ be the filter given by the ideal I generated by finite unions of A_i's. In the permutation model $\mathscr{V}$, $\{A_i : i \in \omega\}$ is a countable family of countable sets without a choice function. To show that $\mathscr{V}$ satisfies $W_{\aleph_0}$, proceed as in Theorem 8.6. If $X \in \mathscr{V}$, $X \in \Delta(E_0)$, then either $X \subseteq \Delta(E_0)$ and X is well-orderable or there is $x \in X$, $x \notin \Delta(E_0)$ and we let $Y = \{\pi x : \pi \in \text{fix}(E_0)\}$. Y has at least two elements and is well-orderable. To show that Y is infinite, argue as follows: otherwise, there is a homomorphism of the symmetric group S_ω of all permutations of ω onto some S_n, and S_ω has a normal subgroup of finite index. This is impossible due to a group-theoretical truth saying that S_ω has only normal subgroups of infinite index.]

5. There is a model of ZF in which the Ultrafilter Theorem (see Section 2.3) fails but every infinite set has a nontrivial ultrafilter.

[*Hint*: In the model $\mathscr{N}$ of Theorem 8.3, with $\aleph_\alpha = \aleph_1$, the Axiom of Choice fails for a family of pairs and so the Prime Ideal Theorem fails. On the other hand, every infinite $X \in \mathscr{N}$ has a countable subset, and since countable subsets of ω are the same in $\mathscr{N}$ as in the ground model $\mathscr{M}$ which satisfies the Axiom of Choice, an ultrafilter on ω in $\mathscr{M}$ yields a nontrivial ultrafilter on X in $\mathscr{N}$.]

8.4. Historical remarks

The generalized version of the Principle of Dependent Choices was formulated by Levy [1964], who also proved Theorem 8.1 and the ZFA versions of Theorems 8.6 and 8.9 and Problem 2. Theorem 8.3 is due to Mostowski [1948] (the ZFA version) and Jech [1966c] (the ZF version).[1] The transfer of W_λ in Theorem 8.6 and of $(\forall\lambda)AC_\lambda$ in Theorem 8.9 is due to Pincus [1969]. Theorems 8.2 and 8.12 were proved by Jensen [1967]. The result in Problem 1 was announced by Gauntt [1967]; the result in Problem 4 is in Pincus [1969].

[1] As for the announcement of Marek [1966], see footnote 2 in Chapter 7.

CHAPTER 9

NONTRANSFERABLE STATEMENTS

9.1. Statements which imply AC in ZF but are weaker than AC in ZFA

Judging from the results of Chapters 5–8, the reader might get the wrong impression that every independence result obtained in ZFA can be transferred into ZF. The transfer was sometimes easier, sometimes harder, but always possible. This is, however, not always the case. In this chapter, we consider several consequences of the Axiom of Choice which are its equivalents in ZF but do not imply it if atoms are permitted.

Let AC denote the Axiom of Choice and consider the following statements:

MC. Axiom of Multiple Choice. *For every family S of nonempty sets, there exists a function f on S such that $f(X)$ is a nonempty finite subset of X for each $X \in S$.*

A. Antichain Principle. *Each partially ordered set has a maximal antichain (i.e., a maximal subset of mutually incomparable elements).*

L. *Every linearly ordered set can be well-ordered.*

P. *The power set of every well-ordered set can be well-ordered.*

We will show that all these statements follow from AC in ZFA, are equivalent to AC in ZF, and AC is independent from each of them in ZFA.

Theorem 9.1

(a). *In* ZFA, AC $\to$ MC $\to$ A $\to$ L $\to$ P.

(b). *In* ZF, P *implies* AC.

Proof. (a). AC $\to$ MC is obvious.

MC $\to$ A: Let $(P, <)$ be a partially ordered set. By MC, there is a function f such that for each nonempty $X \subseteq P$, $f(X)$ is a nonempty finite subset of X. Let g be the following function: for each nonempty $X \subseteq P$, $g(X)$ is the set of all $<$-minimal elements of $f(X)$; $g(X)$ is a nonempty finite antichain. Using g, we construct a maximal antichain in P by transfinite recursion: We let $A_0 = g(P)$, and for each α, $A_\alpha = g(X)$, where X is the set of all $x \in P$ which are incomparable with all $a \in \bigcup \{A_\beta : \beta < \alpha\}$. The union of all A_α's is a maximal antichain in P.

A $\to$ L: Let $(Q, <)$ be linearly ordered. To show that Q can be well-ordered, it suffices to obtain a choice function on the power set of Q. Let P be the set of all pairs (X, x), where X is a subset of Q and $x \in X$; define a partial ordering $\prec$ on P by:

$$(X, x) \prec (Y, y) \text{ if and only if } X = Y \text{ and } x < y.$$

By A, $(P, \prec)$ has a maximal antichain in A. It is easy to see that A defines a choice function on $\mathscr{P}(Q)$.

L $\to$ P: If X is a well-ordered set, then the power set of X can be linearly ordered (lexicographically). By L, $\mathscr{P}(X)$ can be well-ordered.

(b) Assume that the power set of every well-ordered set is well-orderable. To show that AC holds, it suffices to show that for every limit ordinal α, the set V_α of all sets of rank less than α can be well-ordered. Let α be a fixed limit ordinal; we will show that there exists a sequence $\langle W_\beta : \beta < \alpha \rangle$ such that for each $\beta < \alpha$, W_β is a well-ordering of V_β. This will clearly be sufficient. Let κ be the least ordinal (actually a cardinal) such that there is no one-to-one mapping of κ into V_α. By P, $\mathscr{P}(\kappa)$ can be well-ordered; let us fix a well-ordering W of $\mathscr{P}(\kappa)$. We define W_β by recursion on $\beta < \alpha$: $W_0 = 0$, and if β is a limit ordinal, then we define W_β on $V_\beta = \bigcup_{\gamma<\beta} V_\gamma$ in the obvious way from W_γ's, $\gamma < \beta$. If $\beta = \gamma+1$, then $V_\beta = \mathscr{P}(V_\gamma)$. The set V_γ is well-ordered by W_γ and is therefore in a one-to-one correspondence with some ordinal $\lambda < \kappa$. Then we use this one-to-one correspondence and the well-ordering W of $\mathscr{P}(\kappa)$ to get a well-ordering W_β of $\mathscr{P}(V_\gamma)$.

9.2. Independence results in ZFA

In this section we show that none of the implications in Theorem 9.1(a) can be reversed. For each of the implications we present a permutation model in which the implication fails.

THEOREM 9.2. *There are permutation models $\mathscr{V}_1$, $\mathscr{V}_2$, $\mathscr{V}_3$, $\mathscr{V}_4$ such that:*

(i) *in $\mathscr{V}_1$,* MC *holds and* AC *fails;*

(ii) *in $\mathscr{V}_2$,* A *holds and* MC *fails;*

(iii) *in $\mathscr{V}_3$,* L *holds and* A *fails;*

(iv) *in $\mathscr{V}_4$,* P *holds and* L *fails.*

PROOF. First notice that P holds in every permutation model. By (4.2), a set X can be well-ordered in $\mathscr{V}$ if and only if $\text{fix}(X) \in \mathscr{F}$. Thus if X can be well-ordered in $\mathscr{V}$, then $\text{fix}(X) \in \mathscr{F}$, and since $\text{fix}(\mathscr{P}(X)) = \text{fix}(X)$, it follows that $\mathscr{P}(X)$ can be well-ordered in $\mathscr{V}$. Now we present the four models.

(i). Let $\mathscr{V}_1$ be the second Fraenkel model of Section 4.4, where $A = \bigcup_{n=0}^{\infty} P_n$, $P_n = \{a_n, b_n\}$, $\mathscr{G}$ consists of all permutations which preserve each P_n and $\mathscr{F}$ is given by the ideal of finite subsets of A.

The Axiom of Choice fails in $\mathscr{V}_1$. To show that MC holds, we exhibit a function F which to each nonempty $X \in \mathscr{V}_1$ assigns a nonempty finite subset, and F is symmetric; i.e., $F(\pi X) = \pi(F(X))$ for every $\pi \in \mathscr{G}$. For each nonempty $X \in \mathscr{V}_1$, consider the set $o(X) = \{\pi X : \pi \in \mathscr{G}\}$. Choose one X in $o(X)$, choose some $x \in X$, and let

$$F(X) = \{\pi x : \pi \in \text{sym}\,(X)\}. \tag{9.1}$$

It is rather easy to see that if $\pi X = X$, then $\pi(F(X)) = F(X)$. Thus we can define unambiguously

$$F(\rho X) = \rho(F(X))$$

for each $\rho X \in o(X)$. If we do this for every $o(X)$, we get a symmetric function F such that $F(X)$ is a nonempty subset of X for each nonempty X, and always there is an $x \in X$ such that (9.1) holds.

It suffices to show that $F(X)$ is finite. There is an n such that $\pi x = x$ for every $\pi \in \text{fix}\,(P_0 \cup \ldots \cup P_n)$. Consequently, if π and ρ are two permutations in $\mathscr{G}$ and if π and ρ agree on $P_0 \cup \ldots \cup P_n$, then $\pi x = x$. Thus the size of $F(X)$ is at most the number of equivalence classes under the equivalence $\pi \sim \rho$ if and only if $\pi a_0 = \rho a_0, \ldots, \pi a_n = \rho a_n$. Since this is 2^{n+1}, $F(X)$ is finite.

(ii). Let $\mathscr{V}_2$ be the basic Fraenkel model of Section 4.3, where A is countable, $\mathscr{G}$ is the group of all permutations of A and $\mathscr{F}$ is given by the ideal of finite subsets of A.

To show that MC fails in $\mathscr{V}_2$, we show that the power set of A in $\mathscr{V}_2$ does not have a multiple choice function. If f is a function on $\mathscr{P}^{\mathscr{V}}(A)$ and E is a support of f, then $Z = f(A-E)$ is not a nonempty finite subset of $A-E$. This is because for every nonempty finite $X \subseteq A-E$ there exists $\pi \in \mathscr{G}$ such that $\pi \in \text{fix}\,(E)$, $\pi f = f$ and $\pi X \neq X$.

To show that A holds in $\mathscr{V}_2$, it suffices to show that in $\mathscr{V}_2$ every nonempty family P of sets has a maximal $\subseteq$-antichain, i.e., a maximal subfamily Z such that for any $x, y \in Z$, $x \subseteq y$ implies $x = y$. This is sufficient because every partially ordered set $(P, <)$ can be represented by $(P', \subset)$, where P' consists of all sets $\{y: y \leqslant x\}$ for $x \in P$. The key property of $\mathscr{V}_2$ used in the proof is the following lemma:

LEMMA 9.3. *If $x \in \mathscr{V}_2$ and $\pi \in \mathscr{G}$ is such that $x \subseteq \pi x$, then $x = \pi x$.*

PROOF. Let $x \in \mathscr{V}_2$ and $\pi \in \mathscr{G}$, and assume that $x \subseteq \pi x$. We shall find $k > 1$ and $\rho \in \mathscr{G}$ such that $\rho = \rho^{-1}$ and $\rho x = \pi^k x$. Since $x \subseteq \pi x$, we will have $x \subseteq \rho x = \pi^k x$ and $\rho x = \rho^{-1} x \subseteq x$, so that $\pi^k x = x$; consequently, $\pi x = x$.

Let E be a support of x. For every $a \in A$, let

$$c(a) = \{\pi^i a : i = \ldots, -2, -1, 0, 1, 2, \ldots\}$$

and let D be the union of all $c(a)$, $a \in E$, which are finite. D is a finite set and $\pi D = D$, so that π is a permutation on D. Hence if n is a multiple of the size of each $c(a)$, $a \in D$, then $\pi^n \in \operatorname{fix}(D)$. If $a \in E - D$, then eventually π^n is outside E. Thus there exists $k > 1$ such that $\pi^k a = a$ for each $a \in D$ and $\pi^k a \notin E$ whenever $a \in E - D$. We define ρ by $\rho a = \pi^k a$ and $\rho(\pi^k a) = a$ for each $a \in E$, and $\rho a = a$ otherwise. Obviously, $\rho = \rho^{-1}$, and since ρ and π^k agree on E, we have $\rho x = \pi^k x$.

Let $P \in \mathscr{V}_2$ be a nonempty family of sets. We will construct a maximal $\subseteq$-antichain $Z \in \mathscr{V}_2$ in P. Consider the collection of all antichains Z in P such that $\operatorname{sym}(Z) \supseteq \operatorname{sym}(P)$. It is easy to check that this collection satisfies the assumptions of Zorn's Lemma and so there is a maximal Z in this collection. Since $\operatorname{sym}(Z) \supseteq \operatorname{sym}(P)$, we have $Z \in \mathscr{V}_2$. Now it suffices to prove that Z is a maximal antichain $Z \in \mathscr{V}_2$ in P. This will clearly follow from the following lemma:

LEMMA 9.4. *If Z is an antichain in P such that* $\operatorname{sym}(Z) \supseteq \operatorname{sym}(P)$ *and if $z \in P$ is incomparable with each $x \in Z$, then there exists an antichain $Z' \supseteq Z \cup \{z\}$ such that* $\operatorname{sym}(Z') \supseteq \operatorname{sym}(P)$.

PROOF. Let $Z' = Z \cup \{\pi z: \pi \in \operatorname{sym}(P)\}$. Obviously, $\operatorname{sym}(Z') \supseteq \operatorname{sym}(P)$. To show that Z' is an antichain, note first that each πz ($\pi \in \operatorname{sym}(P)$) is incomparable with each $x \in Z$: If, e.g., $\pi z \subseteq x$, then $z = \pi^{-1}\pi z \subseteq \pi^{-1}x$, a contradiction since $\pi^{-1}x \in Z$. The πz's are mutually incomparable, $\pi_1 z \subseteq \pi_2 z$ implies $z \subseteq \pi_1^{-1}\pi_2 z$, which implies $\pi_1 z = \pi_2 z$ by Lemma 9.3. Thus Z' is an antichain.

(iii) Let $\mathscr{V}_3$ be the following permutation model. Let A be countable and let $<$ be a universal homogeneous partial ordering of A (see Section 7.2

for the details). Let $\mathscr{G}$ be the group of all automorphisms of $(A, <)$ and let $\mathscr{F}$ be the normal filter given by the ideal of finite subsets of A. Every $X \in \mathscr{V}_3$ has a finite support $E \subseteq A$. In the proof we will use the following properties of $(A, <)$, established in Section 7.2.

LEMMA 9.5

(a). *If $(E, \prec)$ is a finite partially ordered set, $E_0 \subseteq E$ and if e_0 is an embedding of $(E_0, \prec)$ into $(A, <)$, then there is an embedding e of $(E, \prec)$ into $(A, <)$ which extends e_0.*

(b). *If E_1 and E_2 are finite subsets of A and i is an isomorphism of $(E_1, <)$ and $(E_2, <)$, then i can be extended to an automorphism of $(A, <)$.*

(c). *If both E_1 and E_2 are supports of X, then $E_1 \cap E_2$ is also a support of X.*

To show that A fails in $\mathscr{V}_3$, we show that $(A, <)$ does not have a maximal antichain in $\mathscr{V}_3$. By Lemma 9.5(a), every finite antichain in A can be extended, and so a maximal antichain would have to be infinite. On the other hand, if Z is an antichain and E is a support of Z, then $Z \subseteq E$. This is because if $a \notin E$ then by Lemma 9.5(a), (b), there exist $\pi \in \mathscr{G}$ and $b \in A$ such that $a < b$, $\pi x = y$ and $\pi \in \text{fix}(E)$; thus $a \in Z$ would imply $b \in Z$.

To show that L holds in $\mathscr{V}_3$, note first that by virtue of Lemma 9.5(c), every $X \in \mathscr{V}_3$ has a least support and that the function assigning to each $X \in \mathscr{V}_3$ its least support is in the model. Secondly, we recall that for each finite $E \subseteq A$, the class $\Delta(E)$ of all $X \in \mathscr{V}_3$ supported by E is well-orderable in the model since $\text{fix}(\Delta(E)) \supseteq \text{fix}(E)$.

Let $(Q, \prec)$ be a linearly ordered set in $\mathscr{V}_3$. We will show that $Q \subseteq \Delta(E)$ for some finite E and thus is well-orderable in $\mathscr{V}_3$. Let S be the set of all least supports of elements of Q. We will show first that S can be linearly ordered in $\mathscr{V}_3$ and then that S is finite. Thus we will have $Q \subseteq \Delta(E)$ where $E = \bigcup S$ and we will be done.

For each $E \in S$, let Q_E be the set of all those $x \in Q$ whose least support is E; $\{Q_E : E \in S\}$ is a partition of Q. Divide S into equivalence classes such that $E_1 \sim E_2$ if and only if $(E_1, <)$ and $(E_2, <)$ are isomorphic. For each equivalence class $C_1 \subseteq S$, choose (in the universe) an $E \in C$ and a well-ordering W_E of Q_E; then let $\mathscr{W} = \{\pi W_E : \pi \in \mathscr{G}$, one E in each equivalence class$\}$. The set $\mathscr{W}$ is symmetric and thus in the model. For each $E \in S$, the set

$$\mathscr{W}_E = \{W \in \mathscr{W} : W \text{ is a well-ordering of } Q_E\}$$

is finite. Thus we can assign (in $\mathscr{V}_3$) to each $E \in S$ the finite set

$$T_E = \{z \in Q_E : z \text{ is the } W\text{-least element of } Q_E \text{ for some } W \in \mathscr{W}_E\}$$

and then we let z_E be the $\prec$-least element of T_E. The function which to each $E \in S$ assigns $z_E \in Q$ is one-to-one and this yields a linear ordering of S.

Let $\prec$ be a linear ordering of S and let E_0 be a support of $(S, \prec)$. We show that each $E \in S$ is a subset of E_0. If E is not a subset of E_0, then by Lemma 9.5(a), (b), there exist $\pi \in \mathscr{G}$ and $E' \neq E$ such that $\pi E = E'$, $\pi E' = E$ and $\pi \in \text{fix}(E_0)$. Hence E cannot be in S because that would imply $E' \in S$, and either $E \prec E'$ or $E' \prec E$ would contradict the fact that $\pi(\prec) = \prec$. Thus S is finite.

(iv). Let $\mathscr{V}_4$ be the ordered Mostowski model of Section 4.5. By the remark at the beginning of this proof, P holds in $\mathscr{V}_4$. On the other hand, L fails in $\mathscr{V}_4$ because AC fails and $\mathscr{V}_4$ satisfies the ordering principle.

9.3. Problems

All the statements considered in Chapter 9 are stronger than the statement P. Compare with the following result.

1. Let $\psi(X)$ be a formula which allows only quantifiers of the type $\forall x \in y$, $\exists x \in y$, $\forall x \in \mathscr{P}(y)$ and $\exists x \in \mathscr{P}(y)$. If AC is provable from $(\forall X)\psi(X)$ in ZF, then in ZFA $(\forall X)\psi(X)$ implies P.

[*Hint*: If $(\forall X)\psi(X)$, then the kernel $\mathscr{M}$ satisfies $(\forall X)\psi(X)$. Since $\mathscr{M}$ is a model of ZF, $\mathscr{M}$ satisfies AC. If X is a well-orderable set, then there is a one-to-one mapping of X into $\mathscr{M}$ and consequently, a one-to-one mapping of $\mathscr{P}(X)$ into $\mathscr{M}$. Since AC holds in $\mathscr{M}$, $\mathscr{P}(X)$ can be well-ordered.]

Consider the following two statements:

INJECTION PRINCIPLE: *For every set S and every proper class C there is a one-to-one mapping of S into C.*

PROJECTION PRINCIPLE: *For every set S and every proper class C there is a mapping of C onto S.*

2. In ZFA, the Injection Principle implies the Projection Principle, which in turn implies the Axiom of Choice. In ZF, the Axiom of Choice implies the Injection Principle.

[*Hint*: If the Projection Principle holds, then for every set S there is a mapping of On onto S, and AC follows. In ZF+AC, prove the Injection Principle using $V = \bigcup_{\alpha \in \text{On}} V_\alpha$.]

There is an alternate axiomatization of set theory, which considers classes as objects of the theory. A transitive model of the class-set theory is given by a transitive class (the universe of the model), and a collection of subclasses of that transitive class which serve as the classes of the model. In particular, if $\mathscr{V}$ is a permutation model then the classes of the model are those subclasses of $\mathscr{V}$ which are symmetric, i.e., $C \subseteq \mathscr{V}$ and $\operatorname{sym}(C) = \{\pi \in \mathscr{G}: \pi''C\} \in \mathscr{F}$.

The permutation models can be generalized to the case of a proper class of atoms. Assume that A is a proper class. In that case, every x belongs to some $\mathscr{P}^{\alpha}(D)$ where $\alpha \in \mathrm{On}$ and D is a subset of the class A. A *permutation* π of A is a permutation of some subset D of A (it is understood that $\pi a = a$ for atoms outside D).[1] Consider a *class* $\mathscr{G}$ of permutations of A which is a group, and a *class* I which is an ideal of subsets of A (e.g., the ideal of all finite subsets). We can call x *symmetric* if there is $E \in I$ such that $\pi x = x$ whenever π leaves E pointwise fixed. Then the class $\mathscr{V}$ of all hereditarily symmetric elements is a transitive model, and its classes are the symmetric classes.[2]

3. There is a model with atoms in which the Axiom of Choice holds and the Projection Principle fails.

[*Hint*: Let A be a proper class, let $\mathscr{G}$ be the group of all permutations of A, and let I be the ideal of finite sets; let $\mathscr{V}$ be the model. Every set of atoms in $\mathscr{V}$ is finite, so that every $X \in \mathscr{V}$ has only finitely many atoms in its transitive closure and thus is well-orderable. On the other hand, the class of all atoms cannot be mapped onto ω by a function in $\mathscr{V}$ (i.e., symmetric).]

*4. There is a model with atoms in which the Projection Principle holds and the Injection Principle fails.

[*Hint*: Let A be a proper class and enumerate the atoms by ω-sequences of ordinals. Consider the permutations π of A which satisfy:

$$\text{(9.2)} \qquad \text{if } a, b \in {}^{\omega}\mathrm{On},\ n \in \omega, \text{ and } a|n = b|n, \text{ then } (\pi a)|n = (\pi b)|n.$$

(Let π be a permutation of a subset D of A satisfying (9.2) and then define π outside D by (9.2) if necessary, and by $(\pi a)(n) = a(n)$ otherwise. Let $\mathscr{G}$ be the class of all such π.) For each n, let H_n consist of all those $\pi \in \mathscr{G}$ such that $(\pi a)|n = a|n$ for all $a \in A$. Call X symmetric iff there exist a finite $E \subseteq A$ and $n \in \omega$ such that $\pi X = X$ whenever $\pi \in \operatorname{fix}(E) \cap H_n$. Let $\mathscr{V}$ be the model. In $\mathscr{V}$, every set of atoms is finite and every set has only finitely

[1] More generally, π may be a class but definable from a set (that will be the case in Problem 4).

[2] Notice that the notion of symmetric and hereditarily symmetric sets can be precisely expressed in the language of ZF+atoms, and the key factor is that each set has only a *set* of atoms in its transitive closure.

many atoms in its transitive closure. There is no one-to-one mapping in $\mathscr{V}$ of ω into A. To show that every class in $\mathscr{V}$ can be mapped onto every set, argue as follows: The Axiom of Choice holds in $\mathscr{V}$. It suffices to show that every proper class in $\mathscr{V}$ can be mapped onto On, for On can be mapped onto any (well-ordered) set. The class A can be mapped onto On by an H_1-symmetric map: $F(a) = a(0)$. Similarly, every class $A_s = \{a \in A: a \supseteq s\}$, where s is a finite sequence of ordinals, can be mapped onto On. If C is a proper class in $\mathscr{V}$, consider the class B of all $A \cap \mathrm{TC}(X)$ for $X \in C$ ($\mathrm{TC}(X)$ is the transitive closure). Elements of B are finite sets of atoms and $\bigcup B$ is either finite, in which case C is well-orderable, or contains A_s for some s. Then C is mapped onto finite sets of ordinals, thus onto On.]

9.4. Historical remarks

The material in this chapter appears in the article of Felgner and Jech [1973]. The fact that P implies AC in ZF was observed by Rubin and Rubin [1963]. The antichain principle A was formulated by Kurepa [1953] and the independence of AC from A appears in Halpern [1962, 1972]. The independence of AC from MC is due to Levy [1962]. The observation in Problem 1 is due to Pincus [1969]. Problems 3 and 4 answer a question from the book of Rubin and Rubin [1963]; the first model is due to Felgner, the other to Jech.

CHAPTER 10

MATHEMATICS WITHOUT CHOICE

10.1. Properties of the real line

In Chapter 2, we gave several examples where the Axiom of Choice is used in mathematical proofs. In the present chapter, we will show that the Axiom of Choice is indispensable in those proofs. We will provide examples of models of set theory where the Axiom of Choice fails and so do the theorems which otherwise can be proved with the help of the Axiom of Choice.

Chapter 2 itself gives several examples of mathematical statements which are equivalent either to the Axiom of Choice or the Prime Ideal Theorem. Since neither is provable in ZF, the statements in question are not provable in ZF either. Here we give some additional examples, mostly related to the properties considered in Chapter 2.

THEOREM 10.1. *There is a model of* ZF *which has an infinite set of real numbers without a countable subset.*

This was of course proved in Chapter 5; the basic Cohen model is the one. Here we are rather interested in the consequences. Thus consider the following statement:

(10.1) *There is an infinite set of reals without a countable subset.*

COROLLARY 10.2. *There is a set S of real numbers and a real number $a \notin S$ such that a is in the closure of S but there is no sequence $\{x_n\}_{n=0}^{\infty}$ of elements of S such that $a = \lim_{n\to\infty} x_n$.*

(Compare with Example 2.4.3(a).)

PROOF. Let D be a D-finite infinite set of reals. The set D must have an accumulation point; for if all $d \in D$ are isolated, then let $\{I_n: n = 0, 1, \ldots\}$ be a fixed enumeration of open intervals with rational endpoints and assign to each $d \in D$ the least n such that $I_n \cap D = \{d\}$; this makes D countable. Let a be an accumulation point of D and let $S = D - \{a\}$. Since S is D-finite, every convergent sequence in S is eventually constant.

COROLLARY 10.3. *There is a function f from the reals into the reals and a real number a such that f is not continuous at a, but whenever $\{x_n\}_{n=0}^{\infty}$ is a sequence with $\lim_{n\to\infty} x_n = a$, then $\lim_{n\to\infty} f(x_n) = f(a)$.*

(Compare with Example 2.4.3(b).)

PROOF. Let S and a respectively be the set and the real number from Corollary 10.2. Let f be defined by $f(x) = 1$ if $x \in S$ and $f(x) = 0$ if $x \notin S$. Then $f(a) = 0$, f is not continuous at a and every sequence $\{x_n\}_{n=0}^{\infty}$ converging to a is eventually outside S.

COROLLARY 10.4. *There is a set T of real numbers which is neither closed nor bounded but every sequence of points in T has a convergent subsequence.*

(Compare with Example 2.4.3(c).)

PROOF. Take the set S from Corollary 10.2, and let T be its image under a homeomorphism which transforms the interval $(\inf S, \sup S)$ into $(-\infty, +\infty)$. Every sequence in T has only finitely many distinct terms.

COROLLARY 10.5. *There is a subspace of the real line which is not separable.*

(Compare with Example 2.4.4.)

PROOF. Any D-finite infinite set of reals (with the natural metric) will do.

(Notice that the subspace still has a countable open base.)

THEOREM 10.6. *There is a model of* ZF *in which the set of all real numbers is a union of countably many countable sets.*

PROOF. We will construct a symmetric model of ZF with the required property. Assume that the ground model $\mathcal{M}$ satisfies the Generalized Continuum Hypothesis. Let P be the set of all finite sets p of triples (n, i, α), such that:

(10.2) (i) $n \in \omega$, $i \in \omega$, $\alpha \in \omega_n$;

(ii) $(n, i, \alpha) \in p$ and $(n, i, \beta) \in p$ implies $\alpha = \beta$.

Let p be stronger than q just in case $p \supseteq q$. (There is the following way to look at P: each p is a union of finitely many p_n's, where p_n is a function from a finite subset of ω into ω_n.) Let $B = \mathrm{RO}(P)$ and let $\mathcal{M}^B$ be the Boolean-valued model. Let G be a generic subset of P and let $\mathcal{M}[G]$ be the corresponding generic extension of $\mathcal{M}$.

Let $\mathscr{G}$ be the group of all permutations of $\omega\times\omega$ with the following property:

(10.3) $$\text{if } \pi(n, i) = (m, j), \text{ then } n = m.$$

Thus for each $\pi\in\mathscr{G}$ there exist permutations π_n of ω $(n\in\omega)$ such that

(10.4) $$\pi(n, i) = (n, \pi_n i).$$

Every $\pi\in\mathscr{G}$ induces an automorphism of $(P, <)$ as follows:

(10.5) $$\pi p = \{(n, \pi_n i, \alpha) : (n, i, \alpha)\in p\}.$$

Consequently, π may be considered an automorphism of B. For every $n\in\omega$, let H_n be the group of all $\pi\in\mathscr{G}$ such that π_k is the identity for all $k < n$. Let $\mathscr{F}$ be the filter on $\mathscr{G}$ generated by $\{H_n : n\in\omega\}$. It is easy to verify that $\mathscr{F}$ is a normal filter.

Let $\mathscr{N}$ be the symmetric submodel of $\mathscr{M}[G]$ determined by $\mathscr{F}$. We are going to show that the set of all reals in $\mathscr{N}$ is a countable union of countable sets. Let x be a real (a subset of ω) in $\mathscr{N}$. Then there is a name $\underset{\sim}{x}$ of x such that $\operatorname{dom}(\underset{\sim}{x}) \subseteq \{m : m\in\omega\}$ and $\operatorname{sym}(\underset{\sim}{x}) \supseteq H_n$ for some $n\in\omega$. For each $n\in\omega$, let B_n be the set of all $u\in B$ such that $\pi u = u$ whenever $\pi\in H_n$. B_n is a complete subalgebra of B.

LEMMA 10.7. *If $u\in B_n$, then*

$$u = \sum\{(p : n) : p\in P \text{ and } p\leqslant u\},$$

where

$$p : n = \{(k, i, \alpha)\in p : k < n\}.$$

PROOF. It suffices to show that if $p\leqslant u$, then $p : n\leqslant u$. If it is not so, and if there is some $q\supseteq p : n$ such that $q\leqslant -u$, then one can easily find $\pi\in H_n$ such that πq and p are compatible, which is a contradiction because $\pi u = u$, $p\leqslant u$ and $\pi q\leqslant -u$.

It follows that for every n and every $\pi\in\mathscr{G}$, if $u\in B_n$, then $\pi u\in B_n$, since by Lemma 10.7, B_n consists exactly of all possible sums

(10.6) $$\sum\{p : p\in S\},\quad S\subseteq P,\quad p = p : n \quad \text{for all } p\in S.$$

Every real $x\in\mathscr{N}$ has a name $\underset{\sim}{x}$ such that

(10.7) $$\operatorname{dom}(\underset{\sim}{x}) \subseteq \{\check{m} : m\in\omega\},\qquad \operatorname{rng}(\underset{\sim}{x}) \subseteq B_n \quad \text{for some } n\in\omega.$$

For each n, let S_n be the set of all $\underset{\sim}{x}$ which satisfy (10.7) and let $\underset{\sim}{R}_n\in\mathscr{M}^B$ be as follows:

$$\operatorname{dom}(\underset{\sim}{R}_n) = S_n,\qquad \underset{\sim}{R}_n(\underset{\sim}{x}) = 1 \quad \text{for all } \underset{\sim}{x}\in S_n.$$

By (10.6), if $\underset{\sim}{x} \in S_n$ and $\pi \in \mathscr{G}$, then $\pi\underset{\sim}{x} \in S_n$ and so $\pi\underset{\sim}{R}_n = \underset{\sim}{R}_n$ for each n. Consequently, the function

$$n \mapsto R_n, \tag{10.8}$$

where R_n is the interpretation of $\underset{\sim}{R}_n$, is in the model $\mathscr{N}$. Also, the set $\bigcup \{R_n : n \in \omega\}$ contains all reals in $\mathscr{N}$, so that if we prove that each R_n is countable in $\mathscr{N}$, we will have a decomposition of the reals into countably many countable sets in $\mathscr{N}$.

LEMMA 10.8. $|R_n| = \aleph_{n+1}$ *for every n.*

PROOF. This follows from (10.6) and the fact that $2^{\aleph_n} = \aleph_{n+1}$.

LEMMA 10.9. *Every* $\aleph_n$ *(in* $\mathscr{M}$*) is countable in* $\mathscr{N}$.

PROOF. Let $n \in \omega$. Let $f \in \mathscr{M}[G]$ be a function with the following name

$$\underset{\sim}{f}((\check{i}, \check{\alpha})^B) = u_{ni\alpha} = \sum \{p \in P : (n, i, \alpha) \in p\} \qquad (i \in \omega, \alpha \in \omega_n).$$

Obviously, if $\pi \in H_{n+1}$, then $\pi u_{ni\alpha} = u_{ni\alpha}$ for all $i \in \omega$, $\alpha \in \omega_n$, and so $f \in \mathscr{N}$. By the genericity of G, f is a mapping of ω onto ω_n.

Now it follows by (10.7) that each S_n is countable in $\mathscr{N}$ and so each R_n is countable in $\mathscr{N}$. This proves that in $\mathscr{N}$, the set of all real numbers is a countable union of countable sets.

It follows from Theorem 10.6 that without the Countable Axiom of Choice it is impossible to define satisfactorily Lebesgue measure, or even Borel sets.

Also, in the above model, $\aleph_1$ is a limit of a countable sequence of countable ordinals, see Problem 3. (As a matter of fact, the singularity of $\aleph_1$ follows directly from the decomposition of the reals into countably many countable sets, cf. Problem 2.)

THEOREM 10.10. *There is a model of* ZF *which satisfies the Principle of Dependent Choices and in which every set of real numbers is Lebesgue measurable.*

We will not give the proof of this theorem, which involves a rather complicated forcing construction. Let us only mention that the model is constructed under the assumption of an inaccessible cardinal.

As outlined in Chapter 2, the Principle of Dependent Choices enables to prove all the 'positive' properties of Lebesgue measure. In the model, every set of reals differs from a Borel set by a set of measure zero. Also, every set of reals differs from a Borel set by a meager set, and consequently, every set has the Baire property. For details, we refer the reader to the literature.

10.2. Algebra without choice

Here we present three counterexamples to algebraic theorems discussed in Examples 2.2.2, 2.2.3 and 2.2.5.

THEOREM 10.11. *There is a model in which a vector space exists which has no basis.*

PROOF. We construct a permutation model with the required properties. Then the First Embedding Theorem can be used to get a model of ZF.

Assume that the set of all atoms is countable, and endow A with operations $+$ and $\cdot$ in such a way that A becomes an infinite-dimensional vector space over the field of rational numbers. Let $\mathscr{G}$ be the group of all automorphisms of A and let $\mathscr{F}$ be the filter given by finite subsets of A. Thus x is symmetric just in case $\operatorname{sym}(x) \supseteq \operatorname{fix}(E)$ for some finite $E \subseteq A$.

Let $\mathscr{V}$ be the permutation model. In $\mathscr{V}$, A is an infinite dimensional vector space over the field of rational numbers. However, A does not have a basis in $\mathscr{V}$.

Let $B \in \mathscr{V}$ be a linearly independent subset of A; we show that B is finite. Let E be a finite subset of A such that $\operatorname{sym}(B) \supseteq \operatorname{fix}(E)$. Let $[E]$ be the subspace of A generated by E. Assume that B is infinite; since $[E]$ is finite-dimensional, there exist distinct $x, y \in B$ outside $[E]$. Let π be an automorphism of A which leaves $[E]$ pointwise fixed and $\pi(x) = x+y$ (show that there is such π). We have $\pi \in \operatorname{fix}(E) \subseteq \operatorname{sym}(B)$ but $\pi x \notin B$. Thus B is finite and A has no basis in $\mathscr{V}$.

THEOREM 10.12. *There is a model in which a free group F exists such that not every subgroup of F is free. In particular, there is a free group whose commutator subgroup is not freely generated.*

PROOF. Again, we present a permutation model with the understanding that the result can be transferred to ZF by means of the First Embedding Theorem.

Let $\mathscr{V}$ be the basic Fraenkel model; that is, $\mathscr{G}$ is the group of all permutations of the countable set A of atoms and $\mathscr{F}$ is given by the ideal of finite sets. Let F be the free group whose free generators are the elements of A. We recall that every element of F can be uniquely written in the form

$$a_1^{\pm 1} a_2^{\pm 1} \ldots a_n^{\pm 1},$$

where the a_i are in A and no a appears adjacent to an a^{-1}. Let C be the *commutator subgroup* of F; C is the subgroup generated by the elements of the form

$$xyx^{-1}y^{-1},$$

where $x, y \in F$. We will show that C is not a free group in $\mathscr{V}$.

Contrariwise, assume that there exists a set $Q \subseteq C$ of free generators of C such that $Q \in \mathscr{V}$. There is a finite set $E \subseteq A$ such that $\pi Q = Q$ whenever $\pi \in \text{fix}(E)$. Let u and v be two distinct elements of A, both outside E and let π be the permutation of A defined by $\pi u = v$, $\pi v = u$, and $\pi a = a$ otherwise.

Note that every permutation of A, in particular π, acts as an automorphism on the free group F. Let us consider the following element of C:

$$c = uvu^{-1}v^{-1},$$

and let

$$c = q_1^{\varepsilon_1} \ldots q_n^{\varepsilon_n}$$

be the unique expression of c in terms of the free generators from Q. Since $\pi c = c^{-1}$, we have

$$c^{-1} = \pi(q_1^{\varepsilon_1}) \ldots \pi(q_n^{\varepsilon_n}) = q_n^{-\varepsilon_n} \ldots q_1^{-\varepsilon_1}.$$

We have $\pi Q = Q$, which means that $\pi q \in Q$ for every $q \in Q$, and by the uniqueness of expression in terms of free generators it follows that

$$\pi q_1 = q_n, \quad \ldots, \quad \pi q_n = q_1$$

and

$$\varepsilon_1 = -\varepsilon_n, \quad \ldots, \quad \varepsilon_n = -\varepsilon_1,$$

which implies that $n = 2k$ and that

(10.9) $$c = b \cdot \pi(b^{-1}),$$

where

(10.10) $$b = q_1^{\varepsilon_1} \ldots q_k^{\varepsilon_k}.$$

Let

(10.11) $$b = a_1^{\nu_1} \ldots a_m^{\nu_m}$$

be the unique expression in terms of generators from A. By (10.10), b belongs to C and since C is generated by commutators $xyx^{-1}y^{-1}$, it follows that for every $a \in A$, in particular for u or v, the sum of its exponents in (10.11) is 0. We will derive a contradiction. By (10.9), we have

$$uvu^{-1}v^{-1} = a_1^{\nu_1} \dots a_m^{\nu_m}\pi(a_m^{-\nu_m}) \dots \pi(a_1^{-\nu_1})$$

which necessarily leads to

$$a_1 = u, \quad a_2 = v, \qquad \nu_1 = \nu_2 = 1$$

and

$$\pi a_3 = a_3, \quad \dots, \quad \pi a_m = a_m.$$

But then none of $a_3, \dots, a_m$ is equal to u and the sum of exponents of u in (10.11) is not 0, a contradiction.

THEOREM 10.13. *There is a model in which a field exists which has no algebraic closure.*

PROOF. We shall present a permutation model $\mathscr{V}$. As for the transfer into ZF, we note that the First Embedding Theorem is not directly applicable, but the following argument will do the trick. First we observe that if $F \subseteq C$ are fields and C is algebraic over F, then if λ is a cardinal and C can be mapped onto λ then $F[x]$ can be mapped onto λ. [To see this, let φ map C onto λ and for every $p \in F[x]$, let $\psi(p) = \{\varphi(r): r \text{ is a root of } p\}$. The range of ψ is a family S of finite subsets of λ, and since C is algebraic over F, the union of S is λ. Hence $|S| = \lambda$.]

Now let $\mathscr{V}$ and $F \in \mathscr{V}$ be as constructed below. Let κ be regular, $\kappa > |\mathscr{P}^{\omega}(F[x])|$. We construct a model $\mathscr{N}$ using the First Embedding Theorem. $F[x]$ cannot be mapped onto κ, either in $\mathscr{V}$ or in $\mathscr{N}$. If $C \in \mathscr{N}$ is an algebraic closure of F then (in $\mathscr{N}$) C cannot be mapped onto κ. However, by Problem 7, C would have to be isomorphic to some $C' \in \mathscr{V}$.

Let $\mathscr{V}$ be the basic Fraenkel model; that is, $\mathscr{G}$ is the group of all permutations of the countable set A of atoms and $\mathscr{F}$ is given by the ideal of finite sets. Let $F \in \mathscr{V}$ be any field of characteristic $\neq 2$ which contains A as a subset. (E.g., let F be the field of fractions of the polynomial ring $R[A]$, where R is the field of real numbers.) We will show that in $\mathscr{V}$, no algebraically closed field of characteristic $\neq 2$ contains A as a subset, and hence F does not have an algebraic closure. Let $C \in \mathscr{V}$ be an algebraically closed field of characteristic $\neq 2$. Let i and $-$i be the two quadratic roots of -1 in C. There is a finite $E \subseteq A$ such that every $\pi \in \text{fix}(E)$ preserves C (not only as a set, but it preserves the operations $+$ and $\cdot$ on C as well) and also $\pi(\mathrm{i}) = \mathrm{i}$, $\pi(-\mathrm{i}) = -\mathrm{i}$; notice that each $\pi \in \text{fix}(E)$ acts as an automorphism on C.

Let u and v be two distinct elements of A outside E and let π be the permutation of A defined by $\pi u = v$, $\pi v = u$, $\pi a = a$ otherwise. We will show that the assumption $u \in C$ leads to a contradiction. Assume that $u \in C$; π is an automorphism of C and so if we let $z = (u-v)^{\frac{1}{2}}$ (that is, z is a quadratic root of $u-v$ in C), then we have $\pi z = (\pi u - \pi v)^{\frac{1}{2}} = (v-u)^{\frac{1}{2}} = \pm \mathrm{i} \cdot z$. Both $\pi z = \mathrm{i} \cdot z$ and $\pi z = -\mathrm{i} \cdot z$ leads to a contradiction; e.g., if $\pi z = \mathrm{i} \cdot z$, then

$$\mathrm{i} = \pi \mathrm{i} = \pi \frac{\pi z}{z} = \frac{\pi^2 z}{\pi z} = \frac{z}{\pi z} = \frac{1}{\mathrm{i}}$$

and so $-1 = \mathrm{i}^2 = 1$, a contradiction since C does not have characteristic 2.

10.3. Problems

1. If there is a D-finite infinite set of reals, then there exists a countable family of sets of reals without a choice function.

[*Hint*: Let D be D-finite and consider the family $\{(r, s) \cap D : r, s$ rational numbers$\}$.]

2. If the set of all reals is a countable union of countable sets, then ω_1 is a limit of a countable sequence of countable ordinals.

[*Hint*: Prove, without using the Axiom of Choice, that there is a mapping f of $2^{\aleph_0}$ onto $\aleph_1$. E.g., f can be defined on $\mathscr{P}(Q)$ (Q = the rationals) as follows: $f(X)$ = the order type of X if X is well-ordered, $f(X) = 0$ otherwise.]

3. In the model $\mathscr{N}$ of Theorem 10.6, $(\aleph_1)^{\mathscr{N}} = (\aleph_\omega)^{\mathscr{M}}$.

[*Hint*: $(\aleph_1)^{\mathscr{N}} \geqslant (\aleph_\omega)^{\mathscr{M}}$ by Lemma 10.9. To show that $\aleph_\omega$ is a cardinal in $\mathscr{N}$, prove that there is no $\underline{f} \in \mathrm{HS}$ which is a name for a function from ω onto $\aleph_\omega$: If $\underline{f} \in \mathrm{HS}$, then for some n, sym $(\underline{f}) \supseteq H_n$. Using Lemma 10.7, we would get $\aleph_\omega$ mutually incompatible conditions of the form $p : n$, a contradiction since there are only $\aleph_n$ such conditions.]

4. In ZF, the following statement implies the Axiom of Choice:

(10.12) *For every real vector space V, if S is a subspace of V then there is a subspace S' of V such that $S \cap S' = \{0\}$ and $[S \cup S'] = V$.*

[*Hint*: Prove that (10.12) implies the Axiom of Multiple Choice (see Chapter 9), which implies the Axiom of Choice in ZF. Let $\mathscr{S}$ be a family of nonempty sets; use (10.12) to find a multiple choice function on $\mathscr{S}$. For every $X \in \mathscr{S}$, let $L(X)$ be the vector space of all linear forms

$a_1x_1 + \dots + a_nx_n$ with indeterminates from X and real coefficients; let $L_0(X)$ be the subspace of $L(X)$ of all $a_1x_1 + \dots + a_nx_n$, such that $a_1 + \dots + a_n = 0$. Let V be the weak direct product of all $L(X)$, $X \in \mathscr{S}$ (i.e., all sums $v_1 + \dots + v_n$ with $v_i \in X_i \in \mathscr{S}$) and let V_0 be the weak direct product of all $L_0(X)$, $X \in \mathscr{S}$. By (10.12), there exists $V_0' \subseteq V$ such that $V_0 \cap V_0' = \{0\}$ and $[V_0 \cup V_0'] = V$. For any $x \in X \in \mathscr{S}$, let $x = v + v'$ be the unique decomposition of $x \in V$ into $v \in V_0$ and $v' \in V_0'$. Claim: if $x, y \in X$, $x = v+v'$, $y = w+w'$, then $v' = w'$. (This is because $v' - w' \in V_0 \cap V_0'$.) Thus we have a function which for each $X \in \mathscr{S}$ chooses $w \in V_0'$ such that $x - w \in V_0$ for every $x \in X$. It follows that the linear form w contains at least one $x \in X$ with a nonzero coefficient. Thus let $f(X)$ be the nonempty finite set of all $x \in X$ which have a nonzero coefficient in w.]

*5. There is a model in which a vector space exists which has two bases of different cardinalities.

[*Hint*: First some algebra: Let V_1, V_2 be two 6-dimensional real vector spaces and let $\alpha_1, \beta_1, \gamma_1$ and $\alpha_2, \beta_2, \gamma_2$ be automorphisms of V_1 and V_2, respectively, such that both $\{1, \alpha_1, \beta_1, \gamma_1\}$ and $\{1, \alpha_2, \beta_2, \gamma_2\}$ is the four-group. Then there is an isomorphism ρ between V_1 and V_2 which is invariant under the automorphisms α, β, γ (i.e., $\rho(\alpha_1 x) = \alpha_2(\rho x)$, etc.). Let the set A of atoms be countable, let $A = B_1 \cup B_2$, $B_1 = \bigcup_{j=0}^{\infty} A_{j1}$, $B_2 = \bigcup_{j=0}^{\infty} A_{j2}$, where the A_{ji} are 6-element sets. Consider the following permutations of 6 elements:

$$\begin{aligned} \alpha_1 &= (12)(34)(5)(6), & \alpha_2 &= (12)(34)(5)(6),\\ \beta_1 &= (13)(24)(5)(6), & \beta_2 &= (12)(3)(4)(56),\\ \gamma_1 &= (14)(23)(5)(6), & \gamma_2 &= (1)(2)(34)(56). \end{aligned}$$

For each j, α is the permutation of $A_{j1} \cup A_{j2}$ which acts as α_1 on A_{j1} and as α_2 on A_{j2}; similarly for β and γ. Let $\mathscr{G}$ be the group of all permutations π of A such that for each j, π preserves both A_{j1} and A_{j2} and acts on $A_{j1} \cup A_{j2}$ as either of 1, α, β or γ. Let $\mathscr{F}$ be the filter given by finite supports. Let $\mathscr{V}$ be the permutation model. Let V_1 and V_2 be the vector spaces over the field of real numbers with respective bases B_1 and B_2. For each j, if V_{j1} and V_{j2} are the subspaces generated by A_{j1} and A_{j2}, there exists an isomorphism ρ_j of V_{j1} and V_{j2} invariant under each α, β and γ; thus $\rho_j \in \mathscr{V}$. Hence V_1 and V_2 are isomorphic in the model. However, B_1 contains a countable subset in $\mathscr{V}$ whereas B_2 does not.]

*6. There is a model in which Urysohn's Lemma fails (see Problem 2.26).

[*Hint*: Let the set A of atoms be countable and linearly ordered as the rational numbers. Consider the order topology on A. Let $\mathscr{G}$ be the set of all order-preserving permutations of A. Let I be the ideal of all subsets E of A satisfying (a) E has only finitely many accumulation points, and (b) every infinite subset of E has an accumulation point. Let $\mathscr{F}$ be the filter generated by $\{\text{fix}(E): E \in I\}$. Let $\mathscr{V}$ be the permutation model. Show that in $\mathscr{V}$, A is a T_4-space and every continuous real-valued function on A is constant.]

*7. Let $\mathscr{V}$ be the basic Fraenkel model, let κ be a regular cardinal and let $\mathscr{N}$ be the symmetric model obtained by the First Embedding Theorem (atoms = subsets of $\mathscr{P}(\kappa)$). If $X \in \mathscr{N}$, then (in $\mathscr{N}$) either there exists $Y \in \mathscr{V}$ such that $|X| = |Y|$ or X can be mapped onto κ.

[*Hint*: Follow the proof of Theorem 8.6. Case (i) is handled precisely as there. In Case (ii), use the Support Theorem. E.g., in the simplest case when $e_0 = 0$, let $x \in X$ have nonempty support with the least element z (in the natural ordering of $\mathscr{P}(\kappa)$); let $\tilde{a} \in \tilde{A}$ be such that $z \in \tilde{a}$. Let $\varphi(y)$ be the least element of the support of y in $\tilde{a}$ for each $y \in X$; φ maps X *onto* $\tilde{a}$. (In general, get a mapping of X onto $\tilde{a} - e_0$.) The set $\tilde{a}$ can be mapped onto κ, e.g., $\psi(z) = \min z$ maps $\tilde{a}$ cofinally into κ (use genericity).]

10.4. Historical remarks

The consequences of a D-finite infinite set of reals appear in Sierpiński [1918], Jaegermann [1965] and Jech [1968a]. Theorem 10.6 was proved by Feferman and Levy [1963]. The model in which every set of reals is Lebesgue measurable was constructed by Solovay [1965, 1970]. The permutation models in Theorems 10.11–10.13 and Problems 5 and 6 are due to Läuchli [1963]. For the transfer of Theorems 10.11 and 10.12 and of Problems 5 and 6 into ZF see Jech and Sochor [1966b]. Problem 7 and the transfer of Theorem 10.13 is due to Pincus [1972a].[1] The result in Problem 4 is from Bleicher [1964].

[1] Pincus also pointed out that the transfer of Theorem 10.13 is not immediate from the Embedding Theorem, as claimed in Jech and Sochor [1966b].

CHAPTER 11

CARDINAL NUMBERS IN SET THEORY WITHOUT CHOICE

11.1. Ordering of cardinal numbers

In set theory with the Axiom of Choice, every infinite set is equivalent to an aleph and the alephs are well-ordered. Without the Axiom of Choice, however, one cannot prove that the ordering $|X| \leqslant |Y|$ is a linear ordering; one can only prove that it is a partial ordering. In Section 2.5, we established the following properties of $|X| \leqslant |Y|$:

(i) $|X| \leqslant |X|$;

(ii) if $|X| \leqslant |Y|$ and $|Y| \leqslant |Z|$, then $|X| \leqslant |Z|$;

(iii) if $|X| \leqslant |Y|$ and $|Y| \leqslant |X|$, then $|X| = |Y|$.

Thus $|X| \leqslant |Y|$ is a partial ordering. We will show that every partial ordering can be represented, in a model of ZF, by the ordering $|X| \leqslant |Y|$ of cardinals.

THEOREM 11.1. *Let $\mathcal{M}$ be a model of* ZF+AC. *Let $(I, \preccurlyeq)$ be a partially ordered set in $\mathcal{M}$. Then there is a symmetric extension $\mathcal{N}$ of $\mathcal{M}$ which satisfies the following*:

There exists a family of sets $\{S_i : i \in I\}$ such that for all $i, j \in I$,

$$i \preccurlyeq j \leftrightarrow |S_i| \leqslant |S_j|. \tag{11.1}$$

PROOF. By the First Embedding Theorem, it suffices to construct a permutation model of ZFA which satisfies (11.1) (and assume that $(I, \preccurlyeq)$ is in the kernel). The Embedding Theorem is then applied to embed the permutation model in a symmetric model of ZF which will also satisfy (11.1).

Thus consider set theory with atoms, and let $(I, \preccurlyeq)$ be a partially ordered set in the kernel. Notice that if we assign to each $i \in I$ the set $\{j \in I : j \preccurlyeq i\}$, then we have an embedding of $(I, \preccurlyeq)$ in $(\mathscr{P}(I), \subseteq)$. Hence it suffices to represent the partially ordered set $(\mathscr{P}(I), \subseteq)$ in the permutation model.

Assume that the set A of atoms has cardinality $|A| = |I| \cdot \aleph_0$. Let

$\{a_{in}: i \in I, n \in \omega\}$ be an enumeration of A. For each $p \subseteq I$, let

$$S_p = \{a_{in}: i \in p, n \in \omega\}.$$

We shall construct a permutation model $\mathscr{V}$ such that:

(i) $\{S_p: p \subseteq I\} \in \mathscr{V}$;

(ii) the function h defined by $h(p) = S_p$ for $p \subseteq I$ is in $\mathscr{V}$;

(iii) $p \subseteq q$ if and only if $\mathscr{V} \models |S_p| \leqslant |S_q|$.

Let $\mathscr{G}$ be the group of all permutations π of A such that $\pi(S_{\{i\}}) = S_{\{i\}}$ for each $i \in I$; i.e., if $\pi a_{in} = a_{jm}$, then $i = j$. Let $\mathscr{F}$ be the filter on $\mathscr{G}$ given by the ideal of finite subsets of A. Each $x \in \mathscr{V}$ has a finite support $E \subseteq A$ such that sym $(x) \supseteq$ fix (E).

It is easy to see that sym $(S_p) = \mathscr{G}$ for each $p \subseteq I$ and that sym $(h) = \mathscr{G}$; hence (i) and (ii) hold. If $p \subseteq q$, then $S_p \subseteq S_q$ and so $|S_p| \leqslant |S_q|$ in $\mathscr{V}$, and it remains to show that if $p \nsubseteq q$, then $|S_p| \nleqslant |S_q|$ in $\mathscr{V}$.

Let $p \nsubseteq q$, and assume that there is a one-to-one mapping $g \in \mathscr{V}$ of S_p into S_q. Let $E \subseteq A$ be a finite support of g. Let $i \in p-q$ and let n, m be two distinct numbers such that neither a_{in} nor a_{im} is in E. Let π be the permutation of A defined by $\pi a_{in} = a_{im}$, $\pi a_{im} = a_{in}$, and $\pi a = a$ otherwise. Since $\pi \in$ fix (E), we have $\pi g = g$. Let $a = g(a_{in})$; since $i \notin q$ and $a \in S_q$, it follows that $\pi a = a$. And we have

$$g(a_{im}) = g(\pi a_{in}) = (\pi g)(\pi a_{in}) = \pi a = a,$$

contrary to the assumption that g is one-to-one.

11.2. Definability of cardinal numbers

One can compare sets by their cardinality, i.e., define the relations $|X| \leqslant |Y|$ and $|X| = |Y|$ without defining the symbol $|X|$. In the presence of the Axiom of Choice, every infinite set is equivalent to an aleph, and so one can let $|X|$ be the appropriate aleph. Without the Axiom of Choice, one can still define $|X|$, namely by

$$|X| = \{Y: |Y| = |X| \text{ and } Y \text{ is of least rank}\}. \tag{11.2}$$

This definition also works in ZFA; that is, if the collection of all atoms is a set, for then (11.2) is a set. However, if one admits in the theory ZFA a proper class of atoms, as we did in Chapter 9 (see the remarks preceding Problem 3), then one cannot use (11.2) to define the cardinal number $|X|$ of X. As a matter of fact, there is no way to define $|X|$ as an operation on sets if one admits the existence of a proper class of atoms:

THEOREM 11.2. *There is a permutation model with a proper class of atoms in which there is no function C, defined for all X, with the property*:

(11.3) $$C(X) = C(Y) \text{ if and only if } |X| = |Y|.$$

We recall the convention that classes of a permutation model $\mathscr{V}$ are the symmetric subclasses of $\mathscr{V}$. It is easy to see that if $\mathscr{C}$ is a definable (from a parameter p) operation in $\mathscr{V}$ then the class

$$\begin{aligned} C &= \{(X, \mathscr{C}(X)) : X \in \mathscr{V}\} \\ &= \{(X, Y) \in \mathscr{V} : \mathscr{V} \models \phi(X, Y, p)\} \end{aligned}$$

(where ϕ is the defining formula) is a symmetric subclass of $\mathscr{V}$ because

$$\mathscr{V} \models \phi(\vec{x}) \leftrightarrow \mathscr{V} \models \phi(\pi\vec{x}),$$

for every $x \in \mathscr{V}$ and every $\pi \in \mathscr{G}$ ($\mathscr{G}$ is the group used in the construction of $\mathscr{V}$). Thus Theorem 11.2 shows the impossibility of defining the operation $|X|$ in ZFA with a class of atoms.

PROOF. We deal with a proper class of atoms and so we have to be a little careful when talking about permutations of atoms, extending the permutations to the entire universe, a group of permutations etc. As in Problems 9.3 and 9.4, one can manage if one considers only permutations which are the identity outside a set of atoms.

Let us divide the atoms into a class of pairs,

$$A = \bigcup_{\alpha \in \mathrm{On}} P_\alpha, \qquad P_\alpha = \{a_\alpha, b_\alpha\},$$

and let us consider all permutations π with the properties

(11.4) (i) π moves only a set of atoms;
(ii) if $\pi a_\alpha \in P_\beta$, then $\pi b_\alpha \in P_\beta$; in other words, $\pi P_\alpha = P_\beta$.

We define symmetric sets as follows: X is *symmetric* if there is a finite set E of atoms and a set S of ordinals such that $\pi X = X$ whenever

(11.5) (i) $\pi a = a$ for every $a \in E$;
(ii) $\pi P_\alpha = P_\alpha$ for every $\alpha \in S$.

The model $\mathscr{V}$ consists of all hereditarily symmetric elements (cf. Footnote 2 in Chapter 9).

The conditions (11.4) and (11.5) guarantee that all atoms are in $\mathscr{V}$ and so are the pairs P_α and any sets of those pairs. A simple argument in the spirit of permutation models shows that if T is an infinite set of ordinals, then the set $\bigcup \{P_\alpha: \alpha \in T\}$ cannot be well-ordered in $\mathscr{V}$, and moreover, if T_1, T_2 are disjoint infinite sets of ordinals, then $\bigcup \{P_\alpha: \alpha \in T_1\}$ and $\bigcup \{P_\alpha: \alpha \in T_2\}$ are not equivalent in $\mathscr{V}$. This will be used to show the nonexistence in $\mathscr{V}$ of a function C which would satisfy (11.3).

Let us assume that C is a function from $\mathscr{V}$ to $\mathscr{V}$, that C is a symmetric class and that C satisfies (11.3) in $\mathscr{V}$. There exists a finite set E of atoms and a set S of ordinals such that if π satisfies (11.5), then $C(\pi X) = \pi(C(X))$ for every X. Let T_1 be a countable set of ordinals disjoint from S and such that $X_1 = \bigcup \{P_\alpha: \alpha \in T_1\}$ is disjoint from E. Look at the set Z of all atoms in the transitive closure of the set $C(X_1)$. If $Z \cap X_1 = 0$, let T_2 be a countable set of ordinals disjoint from both S and T_1 and such that $X_2 = \bigcup \{P_\alpha: \alpha \in T_2\}$ is disjoint from both E and Z. Let π be a permutation which maps X_1 onto X_2 and vice versa, and which is the identity otherwise. We have $C(X_2) = C(\pi X_1) = \pi(C(X_1)) = C(X_1)$, contrary to the fact that X_1 and X_2 are not equivalent in $\mathscr{V}$. If $Z \cap X_1 \neq 0$, let $u \in Z \cap X_1$, e.g., $u = a_\alpha$. Let $\beta \neq \alpha$ be such that $P_\beta \cap Z = 0$, and $\beta \notin S$ and $P_\beta \cap E = 0$. Let π be a permutation which exchanges P_α and P_β, and is the identity otherwise. Clearly, X_1 and πX_1 have the same cardinality in $\mathscr{V}$, whereas $\pi(C(X_1)) \neq C(X_1)$ because πa_α is not in the transitive closure of $\pi(C(X_1))$. In any case we have a contradiction and so such a function C does not exist.

One of the reasons why one uses the alephs to define cardinality of sets if the Axiom of Choice holds is that for every α, $\aleph_\alpha = \omega_\alpha = \{\xi: \xi < \omega_\alpha\}$ and so the cardinality of $\aleph_\alpha$ is $\aleph_\alpha$. In other words, the cardinal numbers are representatives of the equivalence classes consisting of sets of the same cardinality.

Without the Axiom of Choice, nothing of that kind is possible. Precisely, we have the following theorem:

THEOREM 11.3. *There is a model of* ZF *in which there is no function C with the following properties*: *for all X and Y*,

$$(11.6) \qquad \text{(i) } C(X) = C(Y) \textit{ if and only if } |X| = |Y|,$$
$$\text{(ii) } |C(X)| = |X|.$$

PROOF. We use the symmetric model obtained from the ordered Mostowski model by applying the First Embedding Theorem. Thus let $\mathscr{V}$ be the Mostowski model of Section 4.5, let A be the set of all atoms in $\mathscr{V}$, and let $\mathscr{M}$ be the kernel of $\mathscr{V}$. Let $\mathscr{N}$ be a symmetric extension of $\mathscr{M}$ in which $\mathscr{V}$ can be embedded so that $(\mathscr{P}^\omega(A))^{\mathscr{N}} = (\mathscr{P}^\omega(A))^{\mathscr{V}}$. The set A is identified with a set of disjoint sets of subsets of some regular cardinal κ; let

$U = \bigcup \{a: a \in A\}$. By the results of Section 6.2, $\mathcal{N}$ is a support model and $\mathcal{N}$ has a one-to-one function F from $\mathcal{N}$ into $I \times J \times \mathrm{On}$, where I is the set of all finite subsets of A and J is the set of all finite subsets of U.

LEMMA 11.4. *The set A has the following properties in $\mathcal{V}$ (and consequently, in $\mathcal{N}$):*

(a). *A does not have a countable subset.*

(b). *Every subset of A is a union of finitely many intervals and points.*

(c). *Every partition of A has only finitely many members with more than one element.*

(d). *If X and Y are subsets of A of the same cardinality, then $X = Y$ up to a finite set (i.e., $(X-Y) \cup (Y-X)$ is finite); more generally, if $X, Y_1, \ldots, Y_k \subseteq A$ and $|X| = |Y_1| + \ldots + |Y_k|$, then $X = \bigcup_{i=1}^{k} Y_i$ up to a finite set.*

(e). *There is no function f defined on natural numbers such that for every n, $f(n) \subseteq A$ and $|f(n)| \geqslant n$.*

PROOF. This is an exercise in permuting.

LEMMA 11.5. *If X and Y are infinite sets in $\mathcal{N}$ and if $X \subseteq U$ and $Y \subseteq A$, then there is no one-to-one function $f \in \mathcal{N}$ of X onto Y.*

PROOF. Let $p \Vdash \underline{f}$ be a one-to-one function of $\underline{X}$ onto $\underline{Y}$; let (E, e) be a support of f, $e \subseteq \underline{U}$. There are $\underline{x} \in \underline{U} - e$, $a \in A$ and $q \leqslant p$ such that $q \Vdash \underline{f}(\underline{x}) = \underline{a}$. It is easy to find a permutation $\pi \in \mathrm{fix}\,(e)$ such that $\pi a = a$, $\pi \underline{x} \neq \underline{x}$ and such that πq and q are compatible. Then $q \cup \pi q \Vdash \underline{f}(\pi \underline{x}) = \underline{a}$, a contradiction because $[\![\pi \underline{x} = \underline{x}]\!] = 0$.

Now assume that $\mathcal{N}$ has a function C which satisfies (11.6). We will derive a contradiction. As a matter of fact, to get the contradiction, it suffices to assume that a function C with properties (11.6) is defined on the set of all subsets of A. We will define a function f on ω such that $f(n) \subseteq A$ and $|f(n)| \geqslant n$ for every n; that will contradict part (e) of Lemma 11.4.

Let n be fixed. Let X be a subset of A such that $|A - X| = n$, and consider the set $C(X)$ (which does not depend on the choice of X). Using the one-to-one function F of $\mathcal{N}$ into $I \times J \times \mathrm{On}$, let $W = F''C(X)$. For every $z \in J \times \mathrm{On}$, let $W_z = \{E \in I: (E, z) \in W\}$. Since $|X| = |W| = \sum \{|W_z|: z \in J \times \mathrm{On}\}$, it follows by Lemma 11.4(c) that all but finitely many W_z's are singletons.

First we show that there are only finitely many singletons among the W_z's. Otherwise, an infinite subset of A is equivalent to an infinite subset Z of $J \times \mathrm{On}$. Then $|Z| = \sum \{|Z_\alpha|: \alpha \in \mathrm{On}\}$, where $Z_\alpha = \{e \in J: (e, \alpha) \in Z\}$, and by the same argument as above, only finitely many Z_α's are not singletons. Also only finitely many Z_α's are singletons because otherwise we would get a countable subset of A. Hence at least one Z_α is infinite (and equivalent to an infinite subset of A). Now we have $|Z_\alpha| = \sum \{|Z_{\alpha i}|: i \in \omega\}$, where $Z_{\alpha i} = \{e \in Z_\alpha: |e| = i\}$, and again it follows that at least one $Z_{\alpha i}$ is infinite. The set U is linearly ordered and so $|Z_{\alpha i}| \leqslant |U^i|$. Hence U^i has an infinite subset S equivalent to an infinite subset of A. Now $|S| = \sum \{|S_x|: x \in U\}$, where $S_x = \{\vec{y} \in U^{i-1}: (x, \vec{y}) \in S\}$. Using Lemma 11.4(c) again, we get an infinite subset of either U of U^{i-1}, equivalent to an infinite subset of A. In any case, we finally get an infinite subset of U equivalent to an infinite subset of A, contrary to Lemma 11.5.

Thus $|X| = |W_{z_1}| + \ldots + |W_{z_l}| + m$, where l and m are natural numbers, $W_{z_1}, \ldots, W_{z_l}$ are subsets of I and $z_1 < \ldots < z_l$ in the ordering of $J \times \mathrm{On}$. By the same argument as in the preceding paragraph, it follows that we can identify each W_{z_j} with a subset of some A^i. Every subset S of A^i can be represented by a finite sequence $T_1, \ldots, T_r$ of subsets of A and A^{i-1} as follows: let $|S| = \sum \{|T_a|: a \in A\}$ where $T_n = \{\vec{b}: (a, \vec{b}) \in S\}$ and use Lemma 11.4(c). When we do this i times with every W_{z_j}, we get a finite sequence $Y_1, \ldots, Y_k$ of subsets of A such that

$$(11.7) \qquad |X| = |Y_1| + \ldots + |Y_k| + m.$$

By Lemma 11.4 (d), the set $Y = \bigcup_{j=1}^k Y_j$ is equal to X up to a finite set and so $E = A - Y$ is finite. On the other hand, we get from (11.7) that $|Y| \leqslant |X|$ and so $|E| \geqslant n$. Thus given n, we have constructed a finite subset E of A with at least n elements (note that this construction is independent of the choice of X and is uniform for every $n \in \omega$). When we let $f(n) = E$, then we get a function which by Lemma 11.4(e) does not exist.

11.3. Arithmetic of cardinal numbers

If the Axiom of Choice is not used, the cardinal arithmetic lacks the simplicity of the cardinal arithmetic with the Axiom of Choice. Many formulas are no longer true, and those that remain true become very often hard to prove. We present in this section several examples to illustrate some typical results in this area.

Throughout this section, we will use the German letters $\mathfrak{p}, \mathfrak{q}, \ldots$ to denote infinite cardinal numbers, and we will deal with the arithmetic operations

$$\mathfrak{p}+\mathfrak{q}, \qquad \mathfrak{p}\cdot\mathfrak{q}, \qquad \mathfrak{p}^{\mathfrak{q}}, \qquad 2^{\mathfrak{p}},$$

which are defined in the standard way (i.e., $|X|+|Y| = |X \cup Y|$ for disjoint X, Y, $|X| \cdot |Y| = |X \times Y|$, $|X|^{|Y|} = |{}^{Y}X|$ and $2^{|X|} = |\mathscr{P}(X)|$). We will use without saying such elementary properties as

$$(\mathfrak{p}^{\mathfrak{q}})^{m} = \mathfrak{p}^{\mathfrak{q} \cdot m} \quad \text{or} \quad (\mathfrak{p}+\mathfrak{q})^2 = \mathfrak{p}^2+2\mathfrak{p}\mathfrak{q}+\mathfrak{q}^2 \quad \text{or} \quad \mathfrak{p}+\mathfrak{q} \leqslant \mathfrak{p} \cdot \mathfrak{q},$$

as well as the properties of alephs like $\aleph_\alpha^2 = \aleph_\alpha$, etc. For every infinite cardinal number $\mathfrak{p}$, let $\aleph(\mathfrak{p})$ be the *Hartogs number* of $\mathfrak{p}$, i.e., the least ordinal which cannot be embedded by a one-to-one mapping in a set of cardinality $\mathfrak{p}$. For every $\mathfrak{p}$, $\aleph(\mathfrak{p})$ is an aleph, viz. the least aleph $\aleph$ such that $\aleph \nleqslant \mathfrak{p}$.

LEMMA 11.6. *If $\mathfrak{p}$ is an infinite cardinal and $\aleph$ is an aleph, and if*

(11.8) $$\mathfrak{p}+\aleph = \mathfrak{p} \cdot \aleph,$$

then either $\mathfrak{p} \geqslant \aleph$ or $\mathfrak{p} \leqslant \aleph$. In particular, if

(11.9) $$\mathfrak{p}+\aleph(\mathfrak{p}) = \mathfrak{p} \cdot \aleph(\mathfrak{p}),$$

then $\mathfrak{p}$ is an aleph.

PROOF. Let $\mathfrak{p} = |P|$ and let A be a well-ordered set such that $\aleph = |A|$. By (11.8), there exist two disjoint sets P_1 and A_1 such that $P \times A = P_1 \cup A_1$ and $|P_1| = \mathfrak{p}$, $|A_1| = \aleph$. Either there exists $p \in P$ such that $(p, a) \in P_1$ for every $a \in A$ and then $\mathfrak{p} \geqslant \aleph$ because $P_1 \supseteq \{(p, a): a \in A\}$. Or, for every $p \in P$ let a_p be the least $a \in A$ such that $(p, a) \in A_1$, and then $\mathfrak{p} \leqslant \aleph$ because $\{(p, a_p): p \in P\} \subseteq A_1$. In the particular case (11.9), $\mathfrak{p} \geqslant \aleph(\mathfrak{p})$ is impossible, and $\mathfrak{p} \leqslant \aleph(\mathfrak{p})$ implies that $\mathfrak{p}$ is an aleph.

We use the above lemma to establish the first result of this section:

THEOREM 11.7. *If $\mathfrak{p}^2 = \mathfrak{p}$ for every infinite cardinal number $\mathfrak{p}$, then the Axiom of Choice holds.*

PROOF. We will show that under the assumption of the theorem, every infinite cardinal is an aleph. To do so, it suffices to show that

$$\mathfrak{p}+\aleph(\mathfrak{p}) = \mathfrak{p} \cdot \aleph(\mathfrak{p}).$$

Since $\mathfrak{p}+\aleph(\mathfrak{p}) \leqslant \mathfrak{p} \cdot \aleph(\mathfrak{p})$, we have only to show that $\mathfrak{p}+\aleph(\mathfrak{p}) \geqslant \mathfrak{p} \cdot \aleph(\mathfrak{p})$. This is proved as follows:

$$\mathfrak{p}+\aleph(\mathfrak{p}) = (\mathfrak{p}+\aleph(\mathfrak{p}))^2 = \mathfrak{p}^2+2\mathfrak{p} \cdot \aleph(\mathfrak{p})+(\aleph(\mathfrak{p}))^2 \geqslant \mathfrak{p} \cdot \aleph(\mathfrak{p}).$$

The next result is of a similar nature:

THEOREM 11.8. *Assume that $\mathfrak{p}^2 = \mathfrak{q}^2$ implies $\mathfrak{p} = \mathfrak{q}$ for all infinite cardinal numbers $\mathfrak{p}$ and $\mathfrak{q}$. Then the Axiom of Choice holds.*

PROOF. Again, we show that every infinite cardinal number is an aleph. Let $\mathfrak{p}$ be an infinite cardinal and let $\mathfrak{q} = \mathfrak{p}^{\aleph_0}$. It is enough to show that $\mathfrak{q}$ is an aleph. First notice that $\mathfrak{q}^2 = \mathfrak{q}$ because

$$\mathfrak{q}^2 = (\mathfrak{p}^{\aleph_0})^2 = \mathfrak{p}^{2\aleph_0} = \mathfrak{p}^{\aleph_0} = \mathfrak{q}.$$

Thus we have

$$(\mathfrak{q} \cdot \aleph(\mathfrak{q}))^2 = \mathfrak{q} \cdot \aleph(\mathfrak{q}). \tag{11.10}$$

Next we show that

$$(\mathfrak{q} + \aleph(\mathfrak{q}))^2 = \mathfrak{q} \cdot \aleph(\mathfrak{q}). \tag{11.11}$$

On the one hand, we have

$$(\mathfrak{q} + \aleph(\mathfrak{q}))^2 = \mathfrak{q}^2 + 2\mathfrak{q} \cdot \aleph(\mathfrak{q}) + (\aleph(\mathfrak{q}))^2 \geqslant \mathfrak{q} \cdot \aleph(\mathfrak{q}),$$

and on the other hand,

$$\begin{aligned}(\mathfrak{q} + \aleph(\mathfrak{q}))^2 &= \mathfrak{q}^2 + 2\mathfrak{q} \cdot \aleph(\mathfrak{q}) + (\aleph(\mathfrak{q}))^2 = \mathfrak{q} + \aleph(\mathfrak{q}) + \mathfrak{q} \cdot \aleph(\mathfrak{q}) \\ &\leqslant \mathfrak{q} \cdot \aleph(\mathfrak{q}) + \mathfrak{q} \cdot \aleph(\mathfrak{q}) = \mathfrak{q} \cdot 2\aleph(\mathfrak{q}) = \mathfrak{q} \cdot \aleph(\mathfrak{q}).\end{aligned}$$

Hence (11.10) and (11.11) give us

$$(\mathfrak{q} + \aleph(\mathfrak{q}))^2 = (\mathfrak{q} \cdot \aleph(\mathfrak{q}))^2$$

and by the assumption of the theorem, we get

$$\mathfrak{q} + \aleph(\mathfrak{q}) = \mathfrak{q} \cdot \aleph(\mathfrak{q})$$

and consequently, $\mathfrak{q}$ is an aleph.

In contrast to the previous theorem, the following cancellation law is provable without the Axiom of Choice.

THEOREM 11.9. *If $\mathfrak{p}$ and $\mathfrak{q}$ are infinite cardinals and if $2\mathfrak{p} = 2\mathfrak{q}$, then $\mathfrak{p} = \mathfrak{q}$.*

PROOF. Let P, Q, M, N be mutually disjoint sets such that $|P| = |M| = \mathfrak{p}$ and $|Q| = |N| = \mathfrak{q}$. We have $|P \cup M| = |Q \cup M|$, and wish to show that $|P| = |Q|$. There is a one-to-one mapping between P and M and a one-to-one mapping between Q and N. We may as well assume that ρ is a one-to-one function with domain $P \cup Q \cup M \cup N$ such that $\rho''P = M$, $\rho''M = P$, $\rho''Q = N$ and $\rho''N = Q$. Similarly, let σ be a one-to-one function such that $\sigma''(P \cup M) = Q \cup N$ and $\sigma''(Q \cup N) = M \cup P$. We define a sequence π_i of one-to-one functions on P, for all integers i, positive and negative:

$$\begin{aligned} \pi_1 &= \sigma, & \pi_{-1} &= \rho, \\ \pi_2 &= \rho \cdot \sigma, & \pi_{-2} &= \sigma \cdot \rho, \\ \pi_3 &= \sigma \cdot \rho \cdot \sigma, & \pi_{-3} &= \rho \cdot \sigma \cdot \rho, \\ \pi_4 &= \rho \cdot \sigma \cdot \rho \cdot \sigma, & \pi_{-4} &= \sigma \cdot \rho \cdot \sigma \cdot \rho, \\ &\vdots & &\vdots \end{aligned}$$

and for every $p \in P$, consider the sequence

$$(11.12) \qquad S_p = \{\ldots, \pi_{-2}(p), \pi_{-1}(p), p, \pi_1(p), \pi_2(p), \ldots\}.$$

If p_1 and p_2 are in P, then S_{p_1} and S_{p_2} are either disjoint or have the same elements, in which case either the sequences are identical or one is the inverse of the other. If we look at the subsequence of S_p formed by elements of P and Q only, we can see that the elements of P and the elements of Q alternate. Our intention is to pair up the elements of P with the elements of Q in each S, and do it uniformly for every S. This will give us a one-to-one function of P onto Q.

For a given S, pair up first all the pairs (p, q) such that $p, q \in S$ and $q = \pi_1(p)$. Remove these pairs, and in the remaining set, pair up all the pairs p, q such that $q = \pi_{-1}(p)$ (actually there are no such p, q). Then pair up the pairs such that $q = \pi_2(p)$ and remove them, then take $q = \pi_{-2}(p)$, etc. After infinitely many steps, either we have managed to pair up all p's with all q's in S, or some elements are still left. Using common sense, one should be able to see that at worst only one p or one q is left unmatched. Then we have to start all over again, but this time, we have a fixed element of S to start with. Assume that this element is a p (if it is a q, then the argument is similar). Then $S = S_p$, and p determines the order of S:

$$S = \{\ldots, \pi_{-2}p, \pi_{-1}p, p, \pi_1 p, \pi_2 p, \ldots\}.$$

In the subsequence of S consisting of elements of $P \cup Q$, we match p with its nearest neighbour on the right which is in Q, and similarly for the rest of $(P \cup Q) \cap S$.

Thus we have constructed a one-to-one mapping of P onto Q and hence $\mathfrak{p} = \mathfrak{q}$.

We will conclude this section with the proof that the Generalized Continuum Hypothesis implies the Axiom of Choice. To start with, we expect the reader to know the diagonal procedure of Cantor showing that

$$2^{\mathfrak{p}} \not\leq \mathfrak{p},$$

for any cardinal number $\mathfrak{p}$ (viz., if f is a mapping of X into $\mathscr{P}(X)$, then the set $\{x \in X: x \notin f(x)\}$ is not in the range of f). This can be improved somewhat as follows:

LEMMA 11.10. *If* $\mathfrak{p} \geqslant 5$, *then* $2^{\mathfrak{p}} \nleqslant \mathfrak{p}^2$.

PROOF. Let $\mathfrak{p}$ be an infinite cardinal, $|X| = \mathfrak{p}$ and let $\aleph = \aleph(\mathfrak{p})$ be the Hartogs number of $\mathfrak{p}$. We shall assume that $2^{\mathfrak{p}} \leqslant \mathfrak{p}^2$, and construct a one-to-one sequence of elements of X, of length $\aleph$, contrary to the definition of $\aleph$.

Let f be a one-to-one mapping of $\mathscr{P}(X)$ into $X \times X$, and for every infinite ordinal α, let f_α be a one-to-one mapping of α onto $\alpha \times \alpha$ (a refinement of the theorem $\aleph_\alpha^2 = \aleph_\alpha$ in Section 2.5 gives a canonical one-to-one correspondence between α and $\alpha \times \alpha$ for every infinite ordinal α). We construct the sequence $\{x_\alpha: \alpha < \aleph\}$ as follows: First choose $x_0, \ldots, x_4$ arbitrarily. If $n \geqslant 5$, let $C_n = \{x_0, \ldots, x_{n-1}\}$. Since $|\mathscr{P}(C_n)| = 2^n > n^2 = |C_n \times C_n|$, there is a subset U of C_n such that $f(U) \notin C_n \times C_n$. Let U be the first such set (i.e., $U = \{x_{n_1}, \ldots, x_{n_k}\}$, where $\{n_1, \ldots, n_k\}$ is first in a given well-ordering of finite sets of numbers) and if $f(U) = (x, y)$, then let $x_n = x$ if $x \notin C_n$ or $x_n = y$ if $x \in C_n$; in any case, $x_n \notin C_n$. If α is an infinite ordinal, $\alpha < \aleph$, let $C_\alpha = \{x_\xi: \xi < \alpha\}$. From f and f_α, we get a one-to-one mapping g of α into $\mathscr{P}(X)$: $g(\xi) = f^{-1}(x_\eta, x_\zeta)$, where $(\eta, \zeta) = f_\alpha(\zeta)$. Let $U = \{x_\xi \in C_\alpha: x_\xi \notin g(\xi)\}$, and let $(x, y) = f(U)$. It follows that $(x, y) \notin C_\alpha \times C_\alpha$, and thus we can let x_α be either x or y.

The *Continuum Hypothesis* is the hypothesis that every infinite set of reals is either countable or else has the power of the continuum. In other words, if $\mathfrak{q}$ is a cardinal such that

$$\aleph_0 \leqslant \mathfrak{q} \leqslant 2^{\aleph_0},$$

then $\mathfrak{q} = \aleph_0$ or $\mathfrak{q} = 2^{\aleph_0}$. The *Generalized Continuum Hypothesis* says that for all infinite cardinals $\mathfrak{p}$ and $\mathfrak{q}$,

$$\text{(11.13)} \qquad \text{if } \mathfrak{p} \leqslant \mathfrak{q} \leqslant 2^{\mathfrak{p}} \quad \text{then} \quad \mathfrak{q} = \mathfrak{p} \quad \text{or} \quad \mathfrak{q} = 2^{\mathfrak{p}}.$$

THEOREM 11.11. *The Generalized Continuum Hypothesis implies the Axiom of Choice.*

PROOF. Let $\mathfrak{p}$ be an infinite cardinal number. We assume that (11.13) holds for every $\mathfrak{q}$, and show that $\mathfrak{p}^2 = \mathfrak{p}$. The result will then follow from Theorem 11.7. First, we claim that (without any assumption)

$$\text{(11.14)} \qquad \mathfrak{p} \leqslant 2\mathfrak{p} < 2^{\mathfrak{p}}.$$

This is because $\mathfrak{p} \leqslant 2\mathfrak{p}$ and $2\mathfrak{p} \leqslant 2^{\mathfrak{p}}$, and since $2\mathfrak{p} \leqslant \mathfrak{p}^2$, we have $2\mathfrak{p} \neq 2^{\mathfrak{p}}$ by Lemma 11.10. Now it follows from (11.13) that

$$2\mathfrak{p} = \mathfrak{p}. \tag{11.15}$$

From (11.15), it follows that $\mathfrak{p}^2 < 2^{\mathfrak{p}}$. This is because

$$\mathfrak{p}^2 \leqslant (2^{\mathfrak{p}})^2 = 2^{2\mathfrak{p}} = 2^{\mathfrak{p}},$$

and by Lemma 11.10, $\mathfrak{p}^2 \neq 2^{\mathfrak{p}}$. Thus we have

$$\mathfrak{p} \leqslant \mathfrak{p}^2 < 2^{\mathfrak{p}},$$

and using (11.13) once more, we get

$$\mathfrak{p}^2 = \mathfrak{p}.$$

11.4. Problems

A cardinal number is a *Dedekind cardinal* if it is the cardinal number of an infinite D-finite set.

1. If $\mathfrak{p}$ and $\mathfrak{q}$ are Dedekind cardinals, then $\mathfrak{p}+\mathfrak{q}$ and $\mathfrak{p}\cdot\mathfrak{q}$ are Dedekind cardinals.

2.(a). A subset of a D-finite set is D-finite.
(b). An image of a D-finite set need not be D-finite.
[*Hint*: In the second Fraenkel model, A is D-finite but is a union of countably many pairs, thus it can be mapped onto ω.]

3. The union of a D-finite family of finite sets need not be D-finite.
[*Hint*: In the second Fraenkel model, $A = \bigcup_{n=0}^{\infty} P_n$. For each $a \in A$, let $S_a = \{a, n\}$, where n is such that $a \in P_n$. Then $\bigcup \{S_a : a \in A\} \supseteq \omega$.]

4. The union of a D-finite family of mutually disjoint D-finite sets is D-finite.

5(a). The set of all finite one-to-one sequences in a D-finite set is D-finite.
(b). The set of all finite subsets of a D-finite set need not be D-finite.
[*Hint*: The second Fraenkel model.]

6. If there is a Dedekind cardinal, then there is a set C of Dedekind cardinals such that $|C| = 2^{\aleph_0}$, any $\mathfrak{p}, \mathfrak{q} \in C$ are comparable, and C is isomorphic (in the ordering of the cardinals) to the real line R.
[*Hint*: There is a family $\{A_x : x \in R\}$ of subsets of ω such that $A_x \subseteq A_y$ if and only if $x \leqslant y$. Let D be an infinite D-finite set, and let S be the set of all finite one-to-one sequences in D. For $x \in R$, let $S_x = \{s \in S : |s| \in A_x\}$; let $C = \{|S_x| : x \in R\}$.]

***7. Assuming that there is a Dedekind cardinal, are there two incomparable Dedekind cardinals?

Let $|X| \leqslant_* |Y|$ mean that Y can be mapped onto X (or $X = 0$). With the Axiom of Choice, $\leqslant_*$ is the same as $\leqslant$. Without the Axiom of Choice, $\leqslant_*$ is transitive, but $\mathfrak{p} \leqslant_* \mathfrak{q}$ and $\mathfrak{q} \leqslant_* \mathfrak{p}$ does not imply $\mathfrak{p} = \mathfrak{q}$:

8. If there is a Dedekind cardinal, then there exist cardinals $\mathfrak{p}$ and $\mathfrak{q}$ such that $\mathfrak{p} < \mathfrak{q}$ and $\mathfrak{p} \geqslant_* \mathfrak{q}$.

[*Hint*: Let D be an infinite D-finite set, let S be the set of all finite one-to-one sequences in D and let T be obtained from S by removing the empty sequence. Then $|T| < |S|$ (because S is D-finite) and $|T| \geqslant_* |S|$ as can be easily seen.]

9. If there is a Dedekind cardinal, then there are two cardinals $\mathfrak{p}$ and $\mathfrak{q}$ such that $\mathfrak{p} \neq \mathfrak{q}$ and $2^{\mathfrak{p}} = 2^{\mathfrak{q}}$.

[*Hint*: Let S and T be as in Problem 8, let $\mathfrak{p} = |S|$, $\mathfrak{q} = |T|$. Show that $|S| \leqslant_* |T|$ implies $|\mathscr{P}(S)| \leqslant |\mathscr{P}(T)|$; hence $2^{\mathfrak{p}} \leqslant 2^{\mathfrak{q}}$. On the other hand, $\mathfrak{q} \leqslant \mathfrak{p}$ implies $2^{\mathfrak{q}} \leqslant 2^{\mathfrak{p}}$ and so $2^{\mathfrak{p}} = 2^{\mathfrak{q}}$.]

10. If there is a Dedekind cardinal, then there is a cardinal $\mathfrak{p}$ such that $2\mathfrak{p} > \mathfrak{p}$.

[*Hint*: Let D be an infinite D-finite set, let $\mathfrak{q} = |D|$; let $\mathfrak{p} = \mathfrak{q} + \aleph_0$. Claim: $2\mathfrak{p} > \mathfrak{p}$. Otherwise, $2\mathfrak{q} + \aleph_0 = \mathfrak{q} + \aleph_0$. Let $|C| = |D|$, $|A| = \aleph_0$, A, C, D disjoint, f a one-to-one mapping of $A \cup C \cup D$ onto $A \cup D$. The image $f''D$ of D is included in D, except for a finite set, and the complement of $f''D$ in D is infinite. That gives a one-to-one mapping of D onto a proper subset of D, a contradiction.]

**11. There is a model in which $2\mathfrak{p} = \mathfrak{p}$ for every infinite cardinal number $\mathfrak{p}$ while the Axiom of Choice fails.

(Compare with Theorem 11.7.)

12. Assume the Axiom of Choice for Finite Sets (ACF). If $\mathfrak{p}$ and $\mathfrak{q}$ are Dedekind cardinals and $\mathfrak{p}^2 = \mathfrak{q}^2$ then $\mathfrak{p} = \mathfrak{q}$.

(Compare with Theorem 11.8.)

[*Hint*: Let $|P| = \mathfrak{p}$, $|Q| = \mathfrak{q}$, let f be a one-to-one mapping of $P \times P$ onto $Q \times Q$. Let I be the set of all finite subsets $E \subseteq P$ such that $f''(E \times E) = F \times F$ for some $F \subseteq Q$. If $E_1, E_2 \in I$ then $E_1 \cap E_2 \in I$. Claim: For every $x \in P$ there is $E \in I$ such that $x \in E$. Otherwise, one can get an ω-sequence $E_1 \subset E_2 \subset E_1 \subset \ldots$ and by ACF a one-to-one sequence $x_1, x_2, x_3, \ldots$ in P, a contradiction. For every $x \in P$, let E_x be the least $E \in I$ such that

$x \in E$. This constitutes a partition of P into finite sets, and Q has a similar partition. For each E_x there are finitely many one-to-one mappings of E_x onto its counterpart in Q; by ACF, we can choose one for each E_x and combine them to get a one-to-one mapping of P onto Q.]

Let $\mathfrak{p}$ and $\mathfrak{q}$ be cardinal numbers, and let $\mathfrak{p} < \mathfrak{q}$. The cardinal $\mathfrak{q}$ is a 1-*successor* of $\mathfrak{p}$ if for every $\mathfrak{m}$,

$$\mathfrak{p} \leqslant \mathfrak{m} \leqslant \mathfrak{q} \quad \text{implies} \quad \mathfrak{m} = \mathfrak{p} \quad \text{or} \quad \mathfrak{m} = \mathfrak{q}.$$

The cardinal $\mathfrak{q}$ is a 2-*successor* of $\mathfrak{p}$ if for every $\mathfrak{m}$,

$$\mathfrak{p} < \mathfrak{m} \quad \text{implies} \quad \mathfrak{q} \leqslant \mathfrak{m}.$$

The cardinal $\mathfrak{q}$ is a 3-*successor* of $\mathfrak{p}$ if for every $\mathfrak{m}$,

$$\mathfrak{m} < \mathfrak{q} \quad \text{implies} \quad \mathfrak{m} \leqslant \mathfrak{p}.$$

13. Every cardinal $\mathfrak{p}$ has a 1-successor.

[*Hint*: Let $\mathfrak{q} = \mathfrak{p} + \aleph(\mathfrak{p})$. Let $\mathfrak{m}$ be such that $\mathfrak{p} \leqslant \mathfrak{m} \leqslant \mathfrak{q}$. There is $\mathfrak{p}_1$ such that $\mathfrak{m} = \mathfrak{p} + \mathfrak{p}_1$, and there are $\mathfrak{r}$ and $\aleph_\alpha$ such that $\mathfrak{m} = \mathfrak{r} + \aleph_\alpha$ and $\mathfrak{r} \leqslant \mathfrak{p}$, $\aleph_\alpha \leqslant \aleph(\mathfrak{p})$.

Case (i): $\aleph_\alpha = \aleph(\mathfrak{p})$. Then $\aleph(\mathfrak{p}) = \aleph_\alpha \leqslant \mathfrak{m} = \mathfrak{p} + \mathfrak{p}_1$, and since $\aleph(\mathfrak{p}) \not\leqslant \mathfrak{p}$, we have $\aleph(\mathfrak{p}) \leqslant \mathfrak{p}_1$, and so $\mathfrak{m} = \mathfrak{p} + \aleph(\mathfrak{p}) = \mathfrak{q}$.

Case (ii): $\aleph_\alpha < \aleph(\mathfrak{p})$. Then $\aleph_\alpha < \mathfrak{p}$, and so $\mathfrak{p} + \aleph_\alpha = \mathfrak{p}$; thus we have $\mathfrak{m} = \mathfrak{r} + \aleph_\alpha \leqslant \mathfrak{p} + \aleph_\alpha = \mathfrak{p}$.]

14. If every cardinal has a 2-successor, then the Axiom of Choice holds.

[*Hint*: Let $\mathfrak{p}$ be an infinite cardinal, and show that $\mathfrak{p}$ is an aleph. Let $\aleph_\alpha = \aleph(\mathfrak{p})$. Since $\aleph_{\alpha+1} > \aleph_\alpha$, it follows that $\aleph_{\alpha+1} \geqslant$ any 2-successor of $\aleph_\alpha$, and hence $\aleph_{\alpha+1}$ is *the* 2-successor of $\aleph_\alpha$. If $\mathfrak{p} + \aleph_\alpha > \aleph_\alpha$, then $\mathfrak{p} + \aleph_\alpha \geqslant \aleph_{\alpha+1}$, a contradiction. Thus $\mathfrak{p} + \aleph_\alpha = \aleph_\alpha$, and so $\mathfrak{p}$ is an aleph.]

**15. It is not provable in set theory without the Axiom of Choice that every cardinal has a 3-successor.

**16. If every cardinal has a 3-successor, then for every infinite $\mathfrak{p}$, $2\mathfrak{p} = \mathfrak{p}$.

17. In the basic Fraenkel model, $\mathfrak{p}^2 \not\leqslant 2^{\mathfrak{p}}$, where $\mathfrak{p} = |A|$.

(Compare with Lemma 11.10.)

18. In the basic Fraenkel model, $\mathfrak{p} < \mathfrak{p}^2 < \mathfrak{p}^3 < \ldots$.

**19. In the ordered Mostowski model, for every $n \in \omega$ there is a cardinal number $\mathfrak{p}$ such that $\mathfrak{p} < \mathfrak{p}^2 < \ldots < \mathfrak{p}^n = \mathfrak{p}^{n+1}$.

The following three problems do not deal with the arithmetic of cardinals, and are stated here for lack of a better place.

20. The following version of Ramsey's Theorem is not provable without the Axiom of Choice:

If A is an infinite set and $[A]^2 = W_1 \cup W_2$ is a partition of $[A]^2 = \{\{a, b\}: a, b \in A, a \neq b\}$ then there is an infinite homogeneous subset $H \subseteq A$.

[*Hint*: The second Fraenkel (or Cohen) model. Let $A = \bigcup_{n=0}^{\infty} P_n$, and let $\{a, b\} \in W_1$ just in case a, b are in the same pair P_n. A homogeneous set would be an infinite choice function.]

21. There is a model of set theory in which there exists a linearly ordered set $(P, <)$ such that $|P| > \aleph_1$ and every initial segment of P is countable.

[*Hint*: It suffices to construct a permutation model. Let A be a disjoint union of $\aleph_1$ countable sets A_α, $\alpha < \omega_1$, and let each A_α be ordered as rational numbers. Let $\mathscr{G}$ be the group of all π such that for each $\alpha < \omega_1$, $\pi A_\alpha = A_\alpha$, and π is order-preserving on A_α. Let $\mathscr{F}$ be generated by $\{H_\gamma: \gamma < \omega_1\}$, where $H_\gamma = \text{fix}\,(\bigcup \{A_\alpha: \alpha < \gamma\})$. In the model, let $P = A \cup \omega_1$ and order P such that A_α precedes A_β if $\alpha < \beta$, and each α precedes A_α and follows all A_β, $\beta < \alpha$.]

A Boolean algebra B is *atomic* if for every $u \in B$, $u \neq 0$, there is an atom a such that $a \leqslant u$. An *atom* is an $a \in B$ such that $a \neq 0$ and there is no x with $0 < x < a$. An atomic Boolean algebra can be identified with a subalgebra of $(\mathscr{P}(A), \subseteq)$, where A is the set of atoms of B.

22. For every atomic Boolean algebra B, there is a symmetric model $\mathscr{N}$ of ZF such that B is isomorphic in $\mathscr{N}$ to the power set $\mathscr{P}^{\mathscr{N}}(A)$ of some set A.

[*Hint*: It suffices to construct a permutation model. Identify A with the atoms of B. Let $\mathscr{G}$ be the group of all π which move only finitely many atoms. Let $\mathscr{F}$ be generated by the groups fix (E), $E \subseteq A$ finite, and sym (U), where $U \subseteq A$ and $U \in B$. To see that $\mathscr{F}$ is normal, note that $\pi \cdot \text{sym}\,(U) \cdot \pi^{-1} \supseteq \text{sym}\,(U) \cap \text{fix}\,\{a: \pi a \neq a\}$. A subset of A is in the model if and only if it is in B.]

For every infinite countable ordinal α, let W_α be the set of all one-to-one mappings of ω onto α. There are three mutually exclusive alternatives for ω_1:

(A). There is a choice function on the family $\mathscr{W} = \{W_\alpha: \omega \leqslant \alpha < \omega_1\}$.
(B). There is no choice function on $\mathscr{W}$ but ω_1 is regular.
(C). ω_1 is singular.

The alternative (C) holds in the model of Theorem 10.6 (cf. Problem 10.3). The alternative (B) holds, e.g., in the model of Theorem 10.10.

An uncountable aleph $\aleph_\alpha$ is *inaccessible* if it is regular and if $2^{\aleph_\beta} < \aleph_\alpha$ for every $\beta < \alpha$.

23. If the alternative (B) holds, then ω_1 is an inaccessible aleph in the constructible universe L.

[*Hint*: Show that ω_1 is a regular limit cardinal in L; since L satisfies the Generalized Continuum Hypothesis, ω_1 is inaccessible in L. Since ω_1 is regular, it is regular in L. If it were a successor of some κ in L, then in L, which satisfies the Axiom of Choice, we could choose for each α, $\kappa \leqslant \alpha < \omega_1$, a one-to-one mapping of κ onto α; combining those with a fixed mapping of ω onto κ, we would be able to get a choice function on $\mathscr{W}$, a contradiction.]

*24. Let κ be an inaccessible cardinal. Let $\mathcal{N}$ be a symmetric model which is constructed in the same fashion as the model of Theorem 10.6, except that one takes κ in place of $\aleph_\omega$. We have $(\aleph_1)^{\mathcal{N}} = \kappa$, and $\mathcal{N}$ satisfies the alternative (B).

[*Hint*: Go through the proof of Theorem 10.6 to find the relevant properties of the model.]

*25. The model constructed in Problem 24 has a linearly ordered set $(P, <)$ with the property described in Problem 21, i.e., $|P| > \aleph_1$ and every initial segment of P is countable.

[*Hint*: In (10.8), we defined countable sets R_n of reals of the model. In the present model, we can similarly define countable sets of reals R_α, $\alpha < \aleph_1$, such that $R = \bigcup \{R_\alpha : \alpha < \aleph_1\}$. Then let $P = R \cup \omega_1$, and order P in the same fashion as was the P of Problem 21.]

***26. Construct a model of ZF in which every uncountable aleph is singular.

11.5. Historical remarks

The representation of partial orderings by cardinal numbers (Theorem 11.1) is a result of Jech [1966a] (independently obtained by Takahashi [1967]). The undefinability of cardinal numbers in set theory with a class of atoms is due to Gauntt [1967] and Levy [1969]. Theorem 11.3 is in Pincus [1969]. Theorems 11.7 and 11.8 were proved by Tarski [1924a]. For Theorem 11.9, see Bernstein [1905] and Sierpiński [1922]. Theorem 11.11

was announced by Lindenbaum and Tarski [1926]; the first to publish a proof was Sierpiński [1947]. The proof presented here (and Lemma 11.10) is due to Specker [1954].

The result in Problem 6 is due to Tarski [1965]; the open Problem 7 is also due to Tarski, and so are the observations in Problems 8 and 9. For Problems 8 and 9, see also Levy [1960] and Läuchli [1963], respectively; for Problem 10, cf. e.g. Halpern and Howard [1970]. The model in Problem 11 was announced by Sageev [1973].[1] Problem 12 is a special case of a general result of Ellentuck about the cancellation laws for Dedekind cardinals; for more details, consult Ellentuck [1965, 1968a–1970]. The three kinds of successors were defined and the theorems about 1- and 2-successors (Problems 13 and 14) proved by Tarski [1954b]. The independence result about 3-successors (Problem 15) is due to Jech [1966b]. Problem 16 is a recent unpublished result of Truss [1973]. Problems 17 and 23 are due to Specker [1957]; the result of Problem 18 is in Ellentuck [1966]. Problem 20 is due to Kleinberg [1969]; the model in Problem 21 was constructed by Morris. For Problem 22, see Węglorz [1969]. The alternatives (A), (B), (C) were formulated by Church [1927]. The model in Problem 24 was constructed by Hájek [1966].

[1] *Added in proof (April 1973):* In a letter to the author, D. Halpern states that he and P. Howard have constructed, independently of Sageev, a permutation model for $2\mathfrak{p} = \mathfrak{p}$.

CHAPTER 12

SOME PROPERTIES CONTRADICTING THE AXIOM OF CHOICE

12.1. Measurability of $\aleph_1$

In this last chapter, we give a brief introduction to a subject which has in recent years attracted the attention of set theorists. We will consider several properties, among them in particular the *Axiom of Determinateness*, which contradict the Axiom of Choice. These properties turn out to be related to so-called large cardinal axioms. The most popular large cardinal is a measurable cardinal.

An uncountable aleph κ is called a *measurable cardinal* if there exists a nontrivial ultrafilter U over κ which is *κ-complete*, i.e. if $\alpha < \kappa$ and $X_\xi \in U$ for each $\xi < \alpha$, then

$$\bigcap \{X_\xi : \xi < \alpha\} \in U.$$

We say that $X \subseteq \kappa$ has *measure* 1 if $X \in U$ and X has *measure* 0 if $X \notin U$. If the Axiom of Choice holds, then a measurable cardinal is inaccessible; this follows immediately from the following lemma:

LEMMA 12.1. *Let κ be a measurable cardinal. Then:*
(a) *κ is regular;*
(b) *for every cardinal $\lambda < \kappa$, we have $\kappa \nleq 2^\lambda$.*

PROOF. (a). By the κ-completeness, a union of less than κ sets of measure 0 has measure 0. Since U is nontrivial, singletons have measure 0, and so every subset of κ of smaller cardinality has measure 0. Since κ has measure 1, it cannot be a union of less than κ smaller sets; thus κ is regular.

(b). Assume that $\lambda < \kappa$ and $\kappa \leqslant 2^\lambda$. We show that κ cannot be measurable. If it is, then there is a subset $S \subseteq \mathscr{P}(\lambda)$ of cardinality κ and a nontrivial κ-complete ultrafilter U over S. We construct a transfinite sequence S_α, $\alpha < \lambda$, of sets of measure 1 as follows: $S_0 = S$, $S_\alpha = \bigcap \{S_\xi : \xi < \alpha\}$

if α is a limit ordinal, and $S_{\alpha+1}$ = either $\{X \in S_\alpha : \alpha \in X\}$ or $\{X \in S_\alpha : \alpha \notin X\}$ whichever has measure 1. By the κ-completeness, the intersection

$$\bigcap \{S_\alpha : \alpha < \lambda\}$$

should have measure 1, but obviously, this intersection has at most one element, a contradiction.

As a matter of fact, measurable cardinals are much larger than inaccessible cardinals: if κ is measurable, then κ is the κth inaccessible (in ZF+AC).

If the Axiom of Choice is not assumed, then a measurable cardinal need not be inaccessible:

THEOREM 12.2. *There is a model of* ZF *in which* $\aleph_1$ *is a measurable cardinal.*

PROOF. We will construct a symmetric model in which $\aleph_1$ is measurable, under the assumption that the ground model has a measurable cardinal. (This assumption is necessary; cf. Problem 1).

Let $\mathcal{M}$ be the ground model (of ZF+AC) and let κ be a measurable cardinal. We will construct a complete Boolean algebra B, a group of automorphisms $\mathcal{G}$, and a normal filter $\mathcal{F}$ such that in the model $\mathcal{N}$ given by B, $\mathcal{G}$ and $\mathcal{F}$ (and a generic ultrafilter G), κ is the least uncountable cardinal and is measurable. To begin with, let us say that a subset $A \subseteq B$ is *symmetric* if fix $(A) \in \mathcal{F}$, where fix $(A) = \{\pi \in \mathcal{G} : \pi u = u$ for all $u \in A\}$.

LEMMA 12.3. *Assume that every symmetric subset* $A \subseteq B$ *has cardinality less than* κ. *Then* κ *is a measurable cardinal in* $\mathcal{N}$.

PROOF. Let U be a nontrivial κ-complete ultrafilter over κ in $\mathcal{M}$. It suffices to show that $\mathcal{N}$ satisfies the following:

(i). If $X \subseteq \kappa$ then either X or $\kappa - X$ has a subset Y such that $Y \in U$.

(ii). If $\alpha < \kappa$ and $\{X_\xi : \xi < \alpha\}$ is a partition of κ, then for some $\xi_0 < \alpha$ there is a $Y \in U$ such that $Y \subseteq X_{\xi_0}$.

(Actually, (ii) implies (i).) We will prove (i) and leave the slightly more general case (ii) to the reader.

Let $X \in \mathcal{N}$, $X \subseteq \kappa$; let $\underline{X}$ be a symmetric name of X. It is easy to see that the set $A = \{[\![\check{\alpha} \in \underline{X}]\!] : \alpha < \kappa\}$ is a symmetric subset of B. By the assumption, we have $|A| < \kappa$. For every $u \in A$, let $X_u = \{\alpha < \kappa : [\![\check{\alpha} \in \underline{X}]\!] = u\}$. Clearly, $\{X_u : u \in A\}$ is a partition of κ (in $\mathcal{M}$) into less than κ parts; hence for some $u_0 \in A$, we have $X_{u_0} \in U$. Now, if $u_0 \in G$, then $X_{u_0} \subseteq X$, and if $u_0 \notin G$, then $X_{u_0} \subseteq \kappa - X$.

We will now construct the model $\mathcal{N}$. Let $(P, <)$ be the set of all finite sequences of ordinals less than κ, and $p \leqslant q$ just in case p extends q. Let $B = \mathrm{RO}(P)$. Let $\mathcal{G}$ be the group of all automorphisms of B induced by the automorphisms of $(P, <)$. We define the filter $\mathcal{F}$ as follows: For every $Z \subseteq P$, $|Z| < \kappa$, let fix $(Z) = \{\pi \in \mathcal{G}: \pi p = p$ for all $p \in Z\}$. We let $\mathcal{F}$ be the filter on $\mathcal{G}$ generated by $\{\text{fix}\,(Z): Z \subseteq P$ and $|Z| < \kappa\}$. $\mathcal{F}$ is a normal filter. Let G be a generic ultrafilter on B and let $\mathcal{N}$ be the symmetric submodel of $\mathcal{M}[G]$ determined by $\mathcal{F}$.

If $\lambda < \kappa$, then λ is countable in $\mathcal{N}$. To see this, let $\underset{\sim}{f}$ be the name such that

$$[\![\underset{\sim}{f}(\check{n}) = \check{\alpha}]\!] = \sum \{p \in P: p(n) = \alpha\},$$

for every $n \in \omega$, $\alpha < \lambda$. Clearly, $\underset{\sim}{f}$ is symmetric, and its interpretation is a mapping of ω onto λ.

Thus if we can show that every symmetric $A \subseteq B$ has cardinality less than κ, then by Lemma 12.3, κ is measurable in $\mathcal{N}$ and we are done. Let A be a symmetric subset of B. There exists $Z \subseteq P$, $|Z| < \kappa$ such that fix $(A) \supseteq$ fix (Z). For every $u \in A$, let $S_u = \{p \in P: p \leqslant u\}$; if $\pi \in$ fix (Z) then $\pi'' S_u = S_u$. Let $\lambda < \kappa$ be a cardinal such that $p \subseteq \omega \times \lambda$ for all $p \in Z$. We claim that if for some $p \in S_u$, $p(n) \geqslant \lambda$ for some n, then there is $q \in S_u$ such that $q(n) = \lambda$ and $q(m) = p(m)$ for $m \neq n$. To see this, let π be the permutation of $\omega \times \kappa$ which interchanges $(n, p(n))$ and (n, λ) and is the identity otherwise; π induces an automorphism of P, and if we let $q = \pi p$, then $q \in S_u$ because $\pi \in$ fix (Z). Thus each of the sets S_u is uniquely determined by its restriction to $\omega \times (\lambda \cup \{\lambda\})$, and consequently, A has cardinality at most 2^λ, which is less than κ.

12.2. Closed unbounded sets and partition properties

Let κ be a regular uncountable aleph. We say that a set $C \subseteq \kappa$ is *closed and unbounded* in κ if

(i) sup $(C \cap \alpha) \in C$ for every $\alpha < \kappa$;

(ii) for each $\beta < \kappa$, there is $\alpha \in C$ such that $\alpha > \beta$.

LEMMA 12.4. (a). *The intersection of less than κ closed unbounded subsets of κ is closed unbounded.*

(b). *The diagonal intersection of a family $\{C_\xi: \xi < \kappa\}$ of closed unbounded subsets of κ is closed unbounded*:

$$\Delta\{C_\alpha: \alpha < \kappa\} = \{\alpha < \kappa: \alpha \in \bigcap_{\xi<\alpha} C_\xi\}.$$

PROOF. (a). We prove, by induction on $\gamma < \kappa$, that the intersection $C = \{C_\xi: \xi < \gamma\}$ of closed unbounded sets is closed unbounded. First we note that the intersection of two closed unbounded sets is closed unbounded: to show that $C_0 \cap C_1$ is unbounded, we construct a sequence $\alpha_0 < \beta_0 < \alpha_1 < \beta_1 < \ldots < \alpha_n < \beta_n < \ldots$ such that $\alpha_n \in C_0$ and $\beta_n \in C_1$; then $\lim \alpha_n = \lim \beta_n \in C_0 \cap C_1$. Hence we may assume, by the induction hypothesis, that $C_0 \supseteq C_1 \supseteq \ldots \supseteq C_\xi \supseteq \ldots$. Obviously, C is closed. To see that C is unbounded, let $\beta < \kappa$. There exists $\alpha_0 \in C_0$, $\alpha_0 > \beta$; similarly, there exists $\alpha_1 \in C_1$, $\alpha_1 > \alpha_0$. In general, there exists $\alpha_\xi \in C_\xi$, exceeding all α_η, $\eta < \xi$. If we let $\alpha = \lim_{\xi \to \gamma} \alpha_\xi$, then $\alpha \in C_\xi$ for every ξ because each C_ξ is closed. Hence $\alpha \in C$.

(b). Let C_ξ, $\xi < \kappa$, be closed unbounded. By (a), we may assume that $C_0 \supseteq C_1 \supseteq \ldots \supseteq C_\xi \supseteq \ldots$. To show that $C = \Delta\{C_\alpha : \alpha < \kappa\}$ is closed, let α be a limit of an increasing sequence $\{\alpha_\eta : \eta < \gamma\}$, with $\alpha_\eta \in C$ for each $\eta < \gamma$. We have to show that $\alpha \in C_\xi$ for each $\xi < \alpha$. If $\xi < \alpha$, then $\alpha = \lim \{\alpha_\eta : \eta < \gamma \text{ and } \alpha_\eta > \xi\}$ is a limit of a sequence in C_ξ, and so $\alpha \in C_\xi$. To show that C is unbounded, let $\beta < \kappa$. We construct a sequence $\alpha_0 < \alpha_1 < \ldots < \alpha_n < \ldots$ such that $\alpha_0 > \beta$ and $\alpha_{n+1} \in C_{\alpha_n}$. If α is the limit of the α_n's, then $\alpha \in C_\xi$ for all $\xi < \alpha$ and so $\alpha \in C$.

Since any two closed unbounded sets intersect in a closed unbounded set, the collection of closed unbounded sets seems to be a good candidate to generate a nontrivial κ-complete ultrafilter over κ. However, if the Axiom of Choice holds, then it is not so because we can construct a set which neither includes nor is disjoint from a closed unbounded set.

Let S be a subset of κ. We say that S is *stationary*, if S has a nonempty intersection with every closed unbounded subset of κ. The nonstationary sets are complements of sets containing a closed unbounded subset. If there is a stationary set which has a stationary complement, then the filter generated by the closed unbounded sets is not an ultrafilter.

THEOREM 12.5. *Let κ be a regular uncountable aleph.*

(a). *The set*

$$S = \{\alpha < \kappa : \operatorname{cf}(\alpha) = \omega\}$$

is stationary.

(b). *If the Axiom of Choice holds, then S is a disjoint union of κ stationary sets.*

It is easy to see that S is stationary. If C is closed unbounded, then the ωth element of C has cofinality ω. Before we prove the rest of Theorem 12.5, we introduce one more notion. Let $X \subseteq \kappa$. A function $f: \kappa \to \kappa$ is *regressive* on X if $f(\alpha) < \alpha$ for all $\alpha \in X$, $\alpha \neq 0$.

LEMMA 12.6. *Assume the Axiom of Choice. If f is a regressive function on a stationary set $S_0 \subseteq \kappa$, then there exists a stationary set $S \subseteq S_0$ and some $\gamma < \kappa$ such that $f(\alpha) = \gamma$ for all $\alpha \in S$.*

PROOF. Contrariwise, assume that for each $\gamma < \kappa$, the set $\{\alpha \in S_0 : f(\alpha) = \gamma\}$ is nonstationary and choose a closed unbounded set C_γ such that for each $\alpha \in S_0 \cap C$, $f(\alpha) \neq \gamma$. Let C be the diagonal intersection of the C_γ's. It follows that $\alpha \in S_0 \cap C$ implies $f(\alpha) \geqslant \alpha$, a contradiction since S_0 is stationary, C is closed unbounded and f is regressive.

PROOF OF THEOREM 12.5(b). For each $\alpha \in S$, let $f_\alpha : \omega \to \alpha$ be an increasing function with $\lim_{n\to\infty} f_\alpha(n) = \alpha$. First we claim that there is n such that for all γ, the set

$$\{\alpha \in S : f_\alpha(n) \geqslant \gamma\} \tag{12.1}$$

is stationary. If not, then for every n there exists some γ_n and a closed unbounded C_n such that $f_\alpha(n) < \gamma_n$ for all $\alpha \in C_n \cap S$. Then let $\gamma = \sup\{\gamma_n : n \in \omega\}$ and $C = \bigcap\{C_n : n \in \omega\}$; the set $C \cap S$ is stationary, hence unbounded, but for every $\alpha \in C \cap S$, $f_\alpha(n) < \gamma$ for all n, which is impossible since $\alpha = \lim_{n\to\infty} f_\alpha(n)$.

Thus, let n be such that for every $\gamma < \kappa$, the set (12.1) is stationary. We let g be a regressive function on S defined as follows:

$$g(\alpha) = f_\alpha(n). \tag{12.2}$$

For each γ, the set $\{\alpha \in S : g(\alpha) \geqslant \gamma\}$ is stationary, and so by Lemma 12.6, for each $\gamma < \kappa$ there exists $\delta \geqslant \gamma$ such that the set

$$S_\delta = \{\alpha \in S : g(\alpha) = \delta\} \tag{12.3}$$

is stationary. Different δ's give disjoint S_δ's and by the regularity of κ we get κ disjoint stationary subsets of S.

Let us look again at the measurable cardinals. Let κ be a measurable cardinal and let U be a nontrivial κ-complete ultrafilter over κ. We say that U is *normal* if whenever X_ξ, $\xi < \kappa$, are sets of measure 1, then the diagonal intersection

$$\Delta\{X_\xi : \xi < \kappa\} = \{\alpha < \kappa : \alpha \in \bigcap_{\xi<\alpha} X_\xi\} \tag{12.4}$$

has measure 1.

LEMMA 12.7. *U is normal if and only if for every regressive function $f : \kappa \to \kappa$ there exists a set X of measure 1 such that f is constant on X.*

PROOF. (a). First assume that U is normal and let f be regressive. If f is not constant on a set of measure 1, then for every $\gamma < \kappa$, the set $X_\gamma = \{\alpha < \kappa : f(\alpha) \neq \gamma\}$ has measure 1, and the diagonal intersection $X = \Delta\{X_\gamma : \gamma < \kappa\}$ has measure 1. However, $f(\alpha) \geqslant \alpha$ for every $\alpha \in X$.

(b). On the other hand, assume that U is not normal and let $X_0 \supseteq X_1 \supseteq \ldots \supseteq X_\xi \supseteq \ldots$ be sets of measure 1 whose diagonal intersection is empty. For every α, let $f(\alpha)$ be the least ξ such that $\alpha \notin X_\xi$. One can verify that f is regressive but not constant on a set of measure 1.

If D is a normal nontrivial κ-complete ultrafilter over κ, let us call D simply a *normal measure* on κ. One can show that if κ is measurable then κ has a normal measure, but the proof requires a small amount of the Axiom of Choice (see Problem 2). Now we show that closed unbounded sets have normal measure 1.

LEMMA 12.8. *If D is a normal measure on κ and if C is a closed unbounded subset of κ, then $C \in D$.*

PROOF. Assuming $\kappa - C \in D$, let us define a regressive function f on $\kappa - C$ as follows: if $\alpha \in \kappa - C$, let $f(\alpha)$ be the largest $\gamma \in C$ such that $\gamma < \alpha$. Since C is closed and $\alpha \notin C, f(\alpha)$ is defined. The function f should be constant on a set of measure 1, which is absurd since C is unbounded.

Now we will consider some partition properties. If A is a set of ordinals, let

$$[A]^2 = \{(\alpha, \beta) : \alpha, \beta \in A \text{ and } \alpha < \beta\}. \qquad (12.5)$$

(Compare with the definition of $[A]^2$ on p. 99.)

Let κ and λ be alephs. The symbol

$$\kappa \to (\lambda)^2,$$

means that whenever $[\kappa]^2 = W_0 \cup W_1$ is a partition of $[\kappa]^2$ into two parts, then there exists a subset $H \subseteq \kappa$ such that $|H| = \lambda$ and H is *homogeneous*, i.e.,

$$[H]^2 \subseteq W_0 \text{ or } [H]^2 \subseteq W_1.$$

The property $\aleph_0 \to (\aleph_0)^2$ is called Ramsey's Theorem and can be proved without the Axiom of Choice. (The proof suggested in Problem 7.1 does use the Axiom of Choice but that can be eliminated.)

An uncountable aleph κ is *weakly compact* if

$$\kappa \to (\kappa)^2.$$

If the Axiom of Choice holds, then a weakly compact cardinal is inaccessible; actually, there is an analogue of Lemma 12.1:

LEMMA 12.9. *Let κ be a weakly compact cardinal. Then:*
(a) *κ is regular;*
(b) *for every cardinal $\lambda < \kappa$, we have $\kappa \not\leq 2^\lambda$.*

PROOF. (a). Assume that κ is a disjoint union $\bigcup \{A_\gamma: \gamma < \lambda\}$ such that $\lambda < \kappa$ and $|A_\gamma| < \kappa$ for each $\gamma < \lambda$. We define a partition as follows: $(\alpha, \beta) \in W_0$ just in case α and β are in the same A_γ. Obviously, this partition does not have a homogeneous set $H \subseteq \kappa$ of cardinality κ.

(b) Assume that $\kappa \leq 2^\lambda$ for some $\lambda < \kappa$ and let λ be the least such λ. Let P be the set of all 0-1 functions on λ, and let $\prec$ be the lexicographical ordering of P. Let S be a subset of P of cardinality κ; let $S = \{f_\alpha: \alpha < \kappa\}$. We define a partition of $[\kappa]^2$ as follows: if $\alpha < \beta$, then $(\alpha, \beta) \in W_0$ just in case $f_\alpha \prec f_\beta$. A homogeneous set H yields an $\prec$-increasing (or decreasing) κ-sequence $\{h_\alpha: \alpha < \kappa\}$ of elements of P. For every $\alpha < \kappa$, let g_α be the common initial segment of h_α and $h_{\alpha+1}$. For each $\alpha < \kappa$, g_α is a 0-1 function on some $\gamma < \lambda$ and since the sequence $\{h_\alpha: \alpha < \kappa\}$ is $\prec$-monotone, the g_α's are pairwise distinct. Thus we have a subset of $\bigcup \{{}^\gamma 2: \gamma < \lambda\}$ of cardinality κ. This is impossible since κ is regular and $|{}^\gamma 2| \not\geq \kappa$, for any $\gamma < \lambda$.

Without the Axiom of Choice, however, even $\aleph_1$ can be weakly compact (see Problem 3).

The measurable cardinals are weakly compact. And let us mention that it has been proved (in the presence of the Axiom of Choice) that a weakly compact cardinal κ is the κth inaccessible cardinal and a measurable cardinal κ is the κth weakly compact cardinal.

THEOREM 12.10. *If κ is a measurable cardinal, then κ is weakly compact.*

PROOF. Let U be a nontrivial κ-complete ultrafilter over κ. Let $[\kappa]^2 = W_0 \cup W_1$ be a partition of $[\kappa]^2$. Let

$$A = \{\alpha < \kappa: X_\alpha \in U\},$$

where

$$X_\alpha = \{\beta < \kappa: \alpha < \beta \text{ and } (\alpha, \beta) \in W_0\}.$$

Assuming that $A \in U$, we construct $H \subseteq \kappa$ of cardinality κ such that $[H]^2 \subseteq W_0$. (Similarly, if $\kappa - A \in U$, one gets H such that $[H]^2 \subseteq W_1$.) Let h_0 be the least element of A, let h_1 be the least element $h > h_0$ such that $h \in A \cap X_{h_0}$, and in general, let h be the least $h >$ all h_β, $\beta < \alpha$, such that

$$h \in A \cap \bigcap \{X_{h_\beta}: \beta < \alpha\}.$$

Since U is κ-complete and κ is regular, h_α is defined for every $\alpha < \kappa$ and it is easy to verify that $[H]^2 \subseteq W_0$.

Now we introduce a generalization of the partition properties considered above. Let A be a set of ordinals and let α be an ordinal. The set $[A]^\alpha$ consists of increasing α-sequences $\{a_0, a_1, \ldots, a_\xi, \ldots\}_{\xi<\alpha}$ of elements of A. If κ, λ are alephs and α an ordinal, then the symbol

$$\kappa \to (\lambda)^\alpha, \tag{12.6}$$

means that whenever $[\kappa]^\alpha = W_0 \cup W_1$ is a partition of $[\kappa]^\alpha$ into two parts, then there exists a subset $H \subseteq \kappa$ such that $|H| = \lambda$ and H is homogeneous, i.e.,

$$[H]^\alpha \subseteq W_0 \quad \text{or} \quad [H]^\alpha \subseteq W_1.$$

If $\alpha \geqslant \omega$, then the partition property (12.6) contradicts the Axiom of Choice:

LEMMA 12.11. *For any κ, $\kappa \to (\aleph_0)^\omega$ contradicts the Axiom of Choice.*

PROOF. If s, t are two increasing ω-sequences of ordinals less than κ, we let $s \sim t$ just in case s and t differ in a finite number of terms. The relation $s \sim t$ is an equivalence relation, and using the Axiom of Choice, we choose a representative $\sigma(s)$ in each equivalence class. Then we define a partition of $[\kappa]^\omega$ as follows: we let $s \in W_0$ just in case s differs from $\sigma(s)$ in an even number of terms. We leave it to the reader to verify that this partition has no infinite homogeneous set.

Without the Axiom of Choice, the situation is somewhat different. On the one hand, it is known that $\aleph_0 \to (\aleph_0)^\omega$ is consistent (relative to an inaccessible cardinal, see Problem 5). On the other hand, the stronger infinitary partition properties are known to be very strong and their consistency is an open problem. In particular, there is a relation between these properties and measurable cardinals. As an example, we present the following theorem:

THEOREM 12.12. (a). *If $\aleph_1 \to (\aleph_1)^\omega$, then every subset of ω_1 either contains or is disjoint from a closed unbounded subset of ω_1.*

(b). *If $\aleph_1 \to (\aleph_1)^{\omega+\omega}$, then the filter generated by closed unbounded subsets of ω_1 is a normal measure on ω_1.*

PROOF. (a). Let A be a subset of ω_1. We define a partition of $[\omega_1]^\omega$ as follows: $p \in [\omega_1]^\omega$ belongs to W_0 just in case $\lim p \in A$. Let $H \subseteq \omega_1$ be an uncountable homogeneous set. We let C be the set of all *limit points* of H, i.e., the set of all α such that $\alpha = \sup (H \cap \alpha)$ (α need not be in H). The set C is obviously closed and unbounded. Now it is clear that if $[H]^\omega \subseteq W_0$, then $C \subseteq A$, and if $[H]^\omega \subseteq W_1$, then $C \subseteq \omega_1 - A$.

(b). Let D be the family of all subsets of ω_1 that contain a closed unbounded subset. By (a), D is an ultrafilter. To show that D is a normal measure, we have to show that:

(12.7) if $X_n \in D$ for each n, then $\bigcap_{n=0}^{\infty} X_n \in D$,

(12.8) D is normal.

The property (12.7) presents some difficulties and we postpone it for a while. Assuming (12.7), it suffices to prove (12.9) to get (12.8):

(12.9) if $f: \omega_1 \to \omega_1$ is regressive, then there is a closed unbounded set C and $\delta < \omega_1$ such that $f(\alpha) \leqslant \delta$ for all $\alpha \in C$.

(Once we have (12.7), one of the sets $\{\alpha : f(\alpha) = \gamma\}$, $\gamma < \delta$, must have measure 1.)

To show (12.9), assume $\aleph_1 \to (\aleph_1)^\omega$, and let f be regressive. For every $p \in [\omega_1]^\omega$, let p_0 be the first term of $p = \{p_0, p_1, \ldots\}$. We define a partition of $[\omega_1]^\omega$ as follows: $p \in W_0$ just in case $f(\lim p) \leqslant p_0$. Let H be a homogeneous set. First we claim that $[H]^\omega \subseteq W_0$. For if $[H]^\omega \subseteq W_1$, then let $p = \{p_0, p_1, \ldots\} \in H$. Since f is regressive, we have $f(\lim p) \leqslant p_n$ for some n, and if we let $q = \{p_n, p_{n+1}, \ldots\}$, we have $q \in [H]^\omega$ and $f(\lim q) \leqslant q_0$. Thus $[H]^\omega \subseteq W_0$, so that $f(\lim p) \leqslant p_0$ for every $p \in [H]^\omega$. Let δ be the least element of H and let C be the set of all limit points of H. C is closed unbounded and $f(\alpha) \leqslant \delta$ for every $\alpha \in C$.

To show (12.7), we consider the following partition property: The symbol

$$\aleph_1 \to (\aleph_1)^\omega_{2^{\aleph_0}},$$

means that whenever $F: [\omega_1]^\omega \to {}^\omega 2$ is a partition of $[\omega_1]^\omega$ into $2^{\aleph_0}$ pieces, then there exists a set $H \subseteq \omega_1$ such that $|H| = \aleph_1$ and $F(p) = F(q)$ for any $p, q \in [H]^\omega$.

LEMMA 12.13. *If* $\aleph_1 \to (\aleph_1)^{\omega+\omega}$, *then* $\aleph_1 \to (\aleph_1)^\omega_{2^{\aleph_0}}$.

PROOF Let $F: [\omega_1]^\omega \to {}^\omega 2$ be a partition of $[\omega_1]^\omega$ into $2^{\aleph_0}$ pieces; we will use the property $\aleph_1 \to [\aleph_1]^{\omega+\omega}$ to find a homogeneous set for F. We define a partition of $[\omega_1]^{\omega+\omega}$ into two pieces W_0 and W_1 as follows: If $p^\frown q$ is an $(\omega+\omega)$-sequence, we let $p^\frown q \in W_0$ just in case $F(p) = F(q)$. Let H be a homogeneous set for this partition. We claim that $[H]^{\omega+\omega} \subseteq W_0$. Otherwise, $F(p) \neq F(q)$ whenever $p^\frown q \in H$; then break H into $\aleph_1$ ω-sequences p_α, $\alpha < \omega_1$, such that $p_\alpha{}^\frown p_\beta \in [H]^{\omega+\omega}$ whenever $\alpha < \beta$. We have $F(p_\alpha) \neq F(p_\beta)$ whenever $\alpha \neq \beta$ and consequently, $\aleph_1 \leqslant 2^{\aleph_0}$. This is impossible by Lemma 12.9 since $\aleph_1$ is weakly compact.

Thus $F(p) = F(q)$ whenever $p^\frown q \in [H]^{\omega+\omega}$. Now, if p and p' are both in $[H]^\omega$, then we easily find q such that both $p^\frown q$ and $p'^\frown q$ are in $[H]^{\omega+\omega}$ and we have $F(p) = F(q) = F(p')$. Hence H is homogeneous for F.

To finish the proof of Theorem 12.12, we use the partition property $\aleph_1 \to (\aleph_1)^\omega_{2^{\aleph_0}}$ to prove (12.7). Let X_n, $n \in \omega$, be subsets of ω_1, each containing a closed unbounded subset. Let $X = \bigcap \{X_n : n \in \omega\}$. We define a partition $F: [\omega_1]^\omega \to {}^\omega 2$ as follows: We let $F(p) = t$, $t = \{t_n\}_{n=0}^\infty \in {}^\omega 2$, where $t_n = 0$ just in case $\lim p \in A_n$. Let $H \in \omega_1$ be homogeneous, $|H| = \aleph_1$. We claim that if $p \in [H]^\omega$, then $F(p) = \{0, 0, \ldots\}$. This is because for each $n \in \omega$, there exists $p \in [H]^\omega$ such that $\lim p \in A_n$. (To see this, let C_n be a closed unbounded subset of A and take any $p \in [H]$ whose limit is in C_n.) Now let C be the set of all limit points of H. It follows that $C \subseteq A_n$ for each n and so A has a closed unbounded subset.

12.3. The Axiom of Determinateness

With each subset A of ${}^\omega\omega$ we associate the following game G_A: Players I and II successively choose natural numbers:

I: $a_0 \quad a_1 \quad a_2 \quad \ldots$,

II: $\quad b_0 \quad b_1 \quad b_2 \quad \ldots$,

and if the resulting sequence $\{a_0, b_0, a_1, b_1, \ldots\}$ is in A, then I wins; otherwise II wins. A *strategy* σ for the player I is a function defined on finite sequences of numbers with values in ω. A strategy σ is a *winning strategy* for I in G_A, if whenever I plays according to σ, i.e. $a_n = \sigma(b_0, \ldots, b_{n-1})$, then I wins, regardless of what II plays. Similarly, we can define a winning strategy for the player II. The game G_A is *determined* if one of the players has a winning strategy.

AXIOM OF DETERMINATENESS. *For every $A \subseteq {}^\omega\omega$, the game G_A is determined.*

In this section, we are going to give a brief outline of some recent results concerning the Axiom of Determinateness. To start with, we show that the Axiom of Determinateness contradicts the Axiom of Choice.

THEOREM 12.14. *If the set of all real numbers can be well-ordered, then there exists $A \subseteq {}^\omega\omega$ such that G_A is not determined.*

PROOF. Let $\aleph_\gamma$ be an aleph such that $2^{\aleph_0} = \aleph_\gamma$. By transfinite recursion, we construct sets $X_\alpha, Y_\alpha \subseteq {}^\omega\omega$, $\alpha < \aleph_\gamma$, such that:

(i) $X_0 \subseteq X_1 \subseteq \ldots \subseteq X_\alpha \subseteq \ldots$, $\quad Y_0 \subseteq Y_1 \subseteq \ldots \subseteq Y_\alpha \subseteq \ldots$;

(ii) $|X_\alpha| \leqslant |\alpha|$, $|Y_\alpha| \leqslant |\alpha|$;

(iii) $X_\alpha \cap Y_\alpha = 0$;

with the intention that $A = \bigcup_{\alpha<\aleph_\gamma} X_\alpha$ is undetermined. Since the number of strategies is $2^{\aleph_0}$, let σ_α, $\alpha < \aleph_\gamma$, be an enumeration of all strategies. Let $\alpha < \aleph_\gamma$, and assume that we have defined X_β, Y_β for all $\beta < \alpha$. If α is a limit ordinal, then we let $X_\alpha = \bigcup_{\beta<\alpha} X_\beta$, $Y_\alpha = \bigcup_{\beta<\alpha} Y_\beta$. If $\alpha = \beta+1$, we construct $X_{\beta+1}$ and $Y_{\beta+1}$ as follows: The set of all sequences $\sigma_\beta[b] = \{a_0, b_0, a_1, b_1, \ldots\}$ which result from a game where the player I uses the strategy σ_β and the player II plays $b_0, b_1, \ldots$ arbitrarily, has cardinality $2^{\aleph_0}$ (the number of all possible sequences $\{b_0, b_1, \ldots\}$ of the moves of the player II). Thus there exists $b \in {}^\omega\omega$ such that $\sigma_\beta[b] \notin X_\beta$; we pick the least such b (in the well-ordering of ${}^\omega\omega$) and let $Y_{\beta+1} = Y_\beta \cup \{\sigma_\beta[b]\}$. Similarly, there exists $a \in {}^\omega\omega$ such that $[a]\sigma_\beta \notin Y_{\beta+1}$, where $[a]\sigma_\beta$ is the game resulting from I playing a and II playing by σ_β. We pick the least such a and let $X_{\beta+1} = X_\beta \cup \{[a]\sigma_\beta\}$.

Finally, when we let $A = \bigcup_{\alpha<\aleph_\gamma} X_\alpha$, we get an undetermined game G_A: each strategy σ is one of the σ_β's and it follows from the construction that σ_β is not a winning strategy for either of the players in the game G_A.

Some choice is possible even with the Axiom of Determinateness. Actually, the axiom implies a weak version of the Axiom of Choice:

LEMMA 12.15. *The Axiom of Determinateness implies that every countable family of nonempty sets of real numbers has a choice function.*

PROOF. We prove that if $\mathscr{X} = \{X_0, X_1, \ldots\}$ is a countable family of nonempty subsets of ${}^\omega\omega$, then $\mathscr{X}$ has a choice function. If I plays $a = \{a_0, a_1, \ldots\}$ and II plays $b = \{b_0, b_1, \ldots\}$, then we let II win just in case $b \in X_{a_0}$. The player I does not have a winning strategy in this game because once I plays a_0, the player II can easily play b such that $b \in X_{a_0}$. Hence II has a winning strategy σ, and we define a choice function f on $\mathscr{X}$ by letting $f(X_n)$ be the play b of II using σ against I playing $\{n, 0, 0, 0, \ldots\}$.

This weak version of the Axiom of Choice has two consequences, which are of some interest now. One is that ω_1 is a regular cardinal, and the other is that Lebesgue measure is countably additive. We mentioned both these consequences in Section 2.4.

One of the remarkable consequences of the Axiom of Determinateness is the following theorem:

THEOREM 12.16. *The Axiom of Determinateness implies that every set of real numbers is Lebesgue measurable.*

PROOF. A nonmeasurable set has different inner and outer measures. In fact, given a nonmeasurable set, one can construct a subset of the interval [0, 1] with inner measure 0 and outer measure 1. Thus to show that every set is measurable, it suffices to show that for every $X \subseteq [0, 1]$, either X contains a subset U of positive measure, or X is disjoint from a set $U \subseteq [0, 1]$ of positive measure.

Let $X \subseteq [0, 1]$. We describe a game which can be coded as a game of the type considered above and thus is determined. Let $r_0, r_1, r_2 \ldots$ be a sequence of positive real numbers such that

$$\tfrac{1}{2} > r_0 > r_1 > r_2 > \ldots, \qquad \sum_{j=0}^{\infty} r_j < \infty. \tag{12.10}$$

Players I and II successively choose closed subsets $S_0, S_1, S_2, \ldots,$ of [0, 1] such that:

(i) $S_0 \supseteq S_1 \supseteq S_2 \supseteq \ldots$;
(ii) S_n is a union of finitely many closed intervals with rational endpoints;
(iii) the diameter of S_n is $\leqslant (\frac{1}{2})^n$;
(iv) the measure of S_n is exactly $r_0 \cdot r_1 \cdot \ldots \cdot r_n$.

The result of this game is a sequence $\{S_0, S_1, \ldots\}$ converging to a single point p. The player I wins if $p \in X$; otherwise, II wins.

The game is determined. We show that if I has a winning strategy, then X contains a subset of positive measure; similarly, if II has a winning strategy, $[0, 1]-X$ has a subset of positive measure.

Let σ be a winning strategy for the player I. Let n be an odd number, and let S be some move of the player I at the stage $n-1$. We claim that there is a finite family $\{T_1, \ldots, T_k\}$ of subsets of S (and satisfying (ii)–(iv)) such that if $S_1, \ldots, S_k$ are such that each S_i is the $(n+1)^{\text{th}}$ move of I after T_i being the nth move of II, then $S_1, \ldots, S_k$ are pairwise disjoint and the measure of $S_1 \cup \ldots \cup S_k$ is at least $1-2r_{n+1}$ times the measure of S (which is $r_0 \cdot \ldots \cdot r_{n-1}$). The sets $T_1, \ldots, T_k$ can easily be constructed by recursion. Let $S_1, \ldots, S_k$ be the corresponding $(n+1)^{\text{th}}$ moves of I and we denote $\{S_1, \ldots, S_k\} = \mathscr{U}_S$.

Now we construct for each n a disjoint family $\mathscr{U}_{2n}$ of closed sets as follows: Let $\mathscr{U}_0 = \{S\}$, where S is the initial move of I by the strategy σ. If $\mathscr{U}_{2n}$ is defined, let $\mathscr{U}_{2n+2} = \bigcup \{\mathscr{U}_S : S \in \mathscr{U}_{2n}\}$. Let

$$U_{2n} = \bigcup \{S : S \in \mathscr{U}_{2n}\}, \tag{12.11}$$

and let

$$U = \bigcap_{n=0}^{\infty} U_n. \tag{12.12}$$

If $S \in \mathscr{U}_{2n}$, then S is the $2n^{\text{th}}$ move of the player I in some game in which I plays according to σ. Thus if $p \in U$, p is a result of such a game, and so $p \in X$. Hence $U \subseteq X$. We will show that U has a positive measure. It is easy to verify that because of the way $\mathscr{U}_{2n}$ was constructed, the measure of each U_{2n} is at least

$$r_0 \cdot (1-2r_2) \cdot \ldots \cdot (1-2r_{2n}). \tag{12.13}$$

By (12.1), the sequence (12.13) converges to a positive number, the measure of U.

We conclude this chapter by presenting the following theorems which suggest a relationship between the Axiom of Determinateness and large-cardinal axioms.

THEOREM 12.17. *The Axiom of Determinateness implies that $\aleph_1$ is a measurable cardinal.*

PROOF. The present proof is based on the theory of *Turing degrees* of functions from ω to ω, a subject belonging to recursion theory, and we will not attempt to go deeply into it but rather state some facts about degrees which we will use.

The basic notion is a relation 'x is *computable* from y', for any $x, y \in {}^{\omega}\omega$. This relation is reflexive and transitive, and so one can define an equivalence by letting x and y have the same *degree* if x is computable from y and y is computable from x. The *degrees* are the equivalence classes. If d, e are degrees, then we let $d \leqslant e$ just in case any $x \in d$ is computable from any $y \in e$. The basic facts about degrees which we will use are the following:

(i). For every $x \in {}^{\omega}\omega$, there are at most countably many $y \in {}^{\omega}\omega$ computable from x.

(ii). If $\{x_0, x_1, \ldots, x_n, \ldots\}$ is a countable subset of ${}^{\omega}\omega$, then there exists $x \in {}^{\omega}\omega$ such that every x_n is computable from x.

(iii). For every $x \in {}^{\omega}\omega$, there exists $y \in {}^{\omega}\omega$ such that x is computable from y but y is not computable from x. The property (i) holds because there are only countably many algorithms to use in the computation. The property (ii) holds because there exists a one-to-one correspondence between ω and $\omega \times \omega$, and so one can construct a single sequence $x \in {}^{\omega}\omega$ which codes the countable sequence of sequences $\{x_0, x_1, \ldots\} \subseteq {}^{\omega}\omega$. The property (iii) is proved by a diagonal argument similar to Cantor's proof of the uncountability of the reals.

LEMMA 12.18. *The Axiom of Determinateness implies that if A is a set of degrees, then there exists a degree d such that either every $e \geqslant d$ is in A or every $e \geqslant d$ is in the complement of A.*

PROOF. Let A' be the set of all $x \in {}^{\omega}\omega$ whose degree is in A. Let G_A be the corresponding game. We claim that if I has a winning strategy, then there exists a degree d such that every $e \geqslant d$ is in A; similarly if II has a winning strategy.

Let σ be a winning strategy for I. Since σ is a function from sequences of numbers into numbers, we may consider σ an element of ${}^{\omega}\omega$ (under a one-to-one correspondence of finite sequences of numbers with ω). Let d be the degree of σ. Let $e \geqslant d$; to show that $e \in A$, it is sufficient to produce a game $\{a_0, b_0, a_1, b_1, \ldots\}$, whose degree is in A and which is won by the player I. Let $b \in {}^{\omega}\omega$ be such that the degree of b is e. The game $\sigma[b]$ in which II plays b and I plays by σ, is a win for I, and has degree e. This is because b is computable from $\sigma[b]$ and $\sigma[b]$ is computable from b and σ, but since $e \geqslant d$, σ is computable from b and so $\sigma[b]$ is computable from b.

Using Lemma 12.18, we can define a countably complete ultrafilter U_1 over the set D of all degrees. If $A \subseteq D$, we let $A \in U_1$ just in case there is some $d \in A$ such that every $e \geqslant d$ is in A. For every $A \subseteq D$, either A or $D - A$ is in U_1. If $\{A_n : n \in \omega\}$ is a countable family of subsets of D, and if $A_n \in U_1$ for each n, then $A = \bigcap_{n=0}^{\infty} A_n$ is in U_1. This is because by Lemma 12.15, we can choose for each n some x_n such that the degree of x_n is in A_n. Then by property (ii), there exists x such that each x_n is computable from x. If we let d be the degree of x, it follows that every $e \geqslant d$ is in each A_n and so $A \in U_1$.

Now we use the fact that there exists a mapping G of ${}^{\omega}\omega$ onto ω_1 and we define a mapping of D into ω_1 as follows:

$$F(d) = \sup \{G(x): \text{the degree of } x \leqslant d\}. \tag{12.14}$$

By property (i), $F(d)$ is a supremum of a countable set of countable ordinals, and since $\aleph_1$ is regular (by Lemma 12.15), $F(d)$ is a countable ordinal. The range of F is an unbounded subset of ω_1.

Finally, we define an ultrafilter on ω_1 as follows:

$$X \in U \text{ if and only if } F_{-1}(X) \in U_1 \tag{12.15}$$

for every $X \subseteq \omega_1$. It is easy to verify that U is a countably complete ultrafilter. To show that U is nontrivial, it suffices to show that for each α,

$$\{d \in D : F(d) \geqslant \alpha\} \in U_1. \tag{12.16}$$

To prove (12.16), let $x \in {}^{\omega}\omega$ be such that $G(x) = \alpha$. By the definition of U_1, the set of all d such that $d \geqslant$ the degree of x is in U_1. By the definition of F, $F(d) \geqslant \alpha$ whenever $d \geqslant$ the degree of x. This proves (12.16).

The last two theorems are stated here without proof. The methods used in the proof are somewhat outside the scope of this book. Notice that Theorem 12.20 implies both Theorems 12.17 and 12.19 by the results of Section 12.2.

THEOREM 12.19. *The Axiom of Determinateness implies that every subset of ω_1 either contains or is disjoint from a closed unbounded subset of ω_1.*

THEOREM 12.20. *The Axiom of Determinateness implies $\aleph_1 \to (\aleph_1)^{\omega_1}$.*

12.4. Problems

*1. If there is a measurable cardinal, then there is a transitive model of ZF+AC which has a measurable cardinal.

[*Hint*: See Problem 3.13. For any set A, $L[A]$ is a model of ZF+AC, and $L[A] = L[\bar{A}]$, where $\bar{A} = A \cap L[A]$ and $\bar{A} \in L[A]$. If U is a nontrivial κ-complete ultrafilter over κ, then $\bar{U}$ is a nontrivial κ-complete ultrafilter over κ in $L[U]$.]

2. Assume the Principle of Dependent Choices. If κ is a measurable cardinal, then κ has a normal measure.

[*Hint*: Let U be a nontrivial κ-complete ultrafilter over κ. For $f, g \in {}^{\kappa}\kappa$ let $f \sim g$ if $\{\alpha : f(\alpha) = f(\alpha)\} \in U$ and $f < g$ if $\{\alpha : f(\alpha) < g(\alpha)\} \in U$. The relation $\sim$ is an equivalence, and $<$ is a linear ordering of the equivalence classes. Since U is closed under countable intersections, there is no descending ω-sequence $f_0 > f_1 > f_2 > \ldots$. By the Principle of Dependent Choices, $<$ is a well-ordering. Let f be the least function such that $f > c_\gamma$ for each γ, where $c_\gamma(\alpha) = \gamma$ for all α. Let $X \in D$ just in case $f_{-1}(X) \in U$. Verify that D is a normal measure.]

**3. If there is a weakly compact cardinal in the ground model, then there is a symmetric extension in which $\aleph_1$ is weakly compact.

**4. If κ is weakly compact, then κ is weakly compact in the constructible universe L.

**5. If there is an inaccessible cardinal in the ground model, then there is a symmetric extension which satisfies $\aleph_0 \to (\aleph_0)^{\omega}$. (In fact, this is true in the model of Theorem 10.10.)

***6. Show that the Axiom of Determinateness implies $\aleph_0 \to (\aleph_0)^\omega$.

Prove the consistency relative to some large cardinal axioms:

***7. For every $A \subseteq \omega_1$, either A or $\omega_1 - A$ has a closed unbounded subset.

***8. $\aleph_1 \to (\aleph_1)^\omega$.

***9. $\aleph_1 \to (\aleph_1)^{\omega_1}$.

***10. The Axiom of Determinateness.

12.5. Historical remarks

Measurable cardinals were introduced by Ulam [1930], who proved that a measurable cardinal is inaccessible. The notion of a normal measure and its construction (Problem 2) is due to Scott [1961]. The construction in Problem 1 is due to Levy. Theorem 12.1 and the result in Problem 3 are in Jech [1968b]. For the results on closed unbounded sets see Fodor [1966] and Solovay [1967a]. Concerning the partition properties, see Erdös and Hajnal [1962] (Theorem 12.10), and Erdös and Rado [1956]. The result in Problem 4 is due to Scott. Theorem 12.12 was proved by Kleinberg [1970]. The result in Problem 5 is due to Mathias [1968]. The formulation and basic consequences of the Axiom of Determinateness can be found in Mycielski [1964b]. Theorem 12.16 was proved by Mycielski and Świerczkowski [1964]. Theorems 12.17 and 12.19 are due to Solovay [1967b]; the present proof of Theorem 12.17 is due to Martin [1968]. Theorem 12.20 was proved by Martin.

APPENDIX

A.1. Equivalents of the Axiom of Choice

Well-ordering Principle	Section 2.1
Zorn's Lemma	Section 2.1
Tukey's Lemma	Section 2.1
Tychonoff's Theorem	Section 2.2, Problem 2.8
Weak Tychonoff's Theorem	Problem 2.9
Prime Ideal Theorem for Lattices	Problem 2.10
$\mathfrak{p}^2 = \mathfrak{p}$	Section 11.3
$\mathfrak{p}^2 = \mathfrak{q}^2 \to \mathfrak{p} = \mathfrak{q}$	Section 11.3

A.2. Equivalents of the Prime Ideal Theorem

Ultrafilter Theorem	Section 2.3
Consistency Principle	Section 2.3
Compactness Theorem	Section 2.3
Stone Representation Theorem	Problem 2.12
Tychonoff's Theorem for H-spaces	Problem 2.14
Compactness of the generalized Cantor space	Problem 2.17
P_3	Problem 2.24

A.3. Various independence results

Independence of the Axiom of Choice	Sections 4.3, 5.3
Independence of C_2	Sections 4.4, 5.4
Prime Ideal Theorem $\nrightarrow$ Axiom of Choice	Section 7.1, Problem 7.6
Selection Principle $\nrightarrow$ Axiom of Choice	Problem 5.21

Ordering Principle $\nrightarrow$ Axiom of Choice	Sections 4.5, 5.5
Ordering Principle $\nrightarrow$ Prime Ideal Theorem	Section 7.2, Problem 7.8
Ordering Principle $\nrightarrow$ Selection Principle	Problem 4.13, Problem 6.11
Selection Principle $\nrightarrow$ Prime Ideal Theorem	Problem 7.8
Prime Ideal Theorem $\nrightarrow$ Selection Principle	Problem 7.6
AC for well-ordered sets (ACW) $\nrightarrow$ Axiom of Choice	Problem 4.14, Problem 5.22
ACW $\nrightarrow$ Ordering Principle	Section 7.3
Ordering Principle $\nrightarrow$ Order Extension Principle	Section 7.2, Problem 7.8
Order Extension Principle $\nrightarrow$ Prime Ideal Theorem	Problem 5.27
Ordering Principle $\nrightarrow$ ACW	Problem 7.12
C_2 $\nrightarrow$ AC for Finite Sets	Section 7.4

AC $\nearrow$ Prime Ideal Theorem $\searrow$ Ordering Principle → AC for Finite Sets → C_2

AC $\searrow$ Selection Principle $\nearrow$ Ordering Principle

ACW ↓ AC for Finite Sets

Independence of $W_{\aleph_0}$	Problem 4.4, Problem 5.18
Principle of Dependent Choices (DC) $\nrightarrow$ Axiom of Choice	Problem 5.26
$DC_{\aleph_1} \nrightarrow (\forall\kappa)AC_\kappa$	Section 8.2
$(\forall\kappa)AC_\kappa \nrightarrow DC_{\aleph_1}$	Section 8.2
$AC_{\aleph_0} \nrightarrow DC$	Section 8.2
$W_{\aleph_0} \nrightarrow AC_{\aleph_0}$	Section 8.2

AC $\nearrow$ $(\forall\kappa)AC_\kappa$

AC $\searrow$ $DC_{\aleph_1}$ $\nearrow$ DC → Countable Axiom of Choice ($AC_{\aleph_0}$) → $W_{\aleph_0}$

Hahn-Banach Theorem $\nrightarrow$ Prime Ideal Theorem	Problem 7.25
Weak Ultrafilter Theorem $\nrightarrow$ Prime Ideal Theorem	Problem 8.5

A.4. Miscellaneous examples

Dedekind-finite set	Problem 4.4, Problem 5.18
Amorphous set	Problem 4.7, Problem 6.5
ω-tree with no infinite chains and antichains	Problem 7.15
Vector space without a basis	Section 10.2, Problem 6.10, Problem 6.16
Free group with a subgroup that is not free	Section 10.2
Field without an algebraic closure	Section 10.2
No nontrivial ultrafilter over ω	Problem 5.24
ω_1 is singular	Section 10.1
The reals as a union of $\aleph_0$ countable sets	Section 10.1
ω_1 is measurable	Section 12.1

REFERENCES

S. Banach and A. Tarski

[1924] Sur la décomposition des ensembles de points en parties respectivement congruentes, *Fund. Math.* **6**, 244–277.

P. Bernays

[1942] A system of axiomatic set theory III, *J. Symb. Logic* **7**, 65–89.

F. Bernstein

[1905] Untersuchungen aus der Mengenlehre, *Math. Ann.* **61**, 117–155.

M. N. Bleicher

[1964] Some theorems on vector spaces and the axiom of choice, *Fund. Math.* **54**, 95–107.

[1965] Multiple choice axioms and axioms of choice for finite sets, *Fund. Math.* **57**, 247–252.

A. Church

[1927] Alternatives to Zermelo's assumption, *Trans. Am. Math. Soc.* **29**, 178–208.

P. J. Cohen

[1963] Independence results in set theory, *The Theory of Models*, J. Addison, L. Henkin and A. Tarski, eds. (North-Holland, Amsterdam, 1965) 39–54.

[1963, 1964] The independence of the continuum hypothesis, I, II, *Proc. Natl. Acad. Sci. U.S.A.* **50**, 1143–1148; **51**, 105–110.

[1966] *Set Theory and the Continuum Hypothesis* (Benjamin, New York).

J. Derrick and F. R. Drake

[1965] Independence of the axiom of choice from variants of the generalized continuum hypothesis, *Proc. Leicester Colloq. Math. Logic*, 75–84.

R. Doss

[1945] Note on two theorems of Mostowski, *J. Symb. Logic* **10**, 13–15.

W. B. Easton

[1964, 1970] *Powers of regular cardinals*, Thesis, Princeton University; *Ann. Math. Logic* **1**, 139–178.

E. Ellentuck

[1965] The universal properties of Dedekind finite cardinals, *Ann. Math.* **82**, 225–248.

[1966] Generalized idempotence in cardinal arithmetic, *Fund. Math.* **58**, 241–258.

[1968a] The representation of cardinals in models of set theory, *Z. Math. Logik u. Grundl. Math.* **14**, 143–158.

[1968b] The first order properties of Dedekind finite integers, *Fund. Math.* **63**, 7–25.

[1969] A choice free theory of Dedekind cardinals, *J. Symb. Logic* **34**, 70–84.
[1970] Almost combinatorial Skolem functions, *J. Symb. Logic* **35**, 65–72.

P. Erdös and A. Hajnal
[1962] Some remarks concerning our paper "On the structure of set mappings", *Acta Math. Acad. Sci. Hung.* **13**, 223–226.

P. Erdös and R. Rado
[1956] A partition calculus in set theory, *Bull. Am. Math. Soc.* **62**, 195–228.

S. Feferman
[1964] Independence of the axiom of choice from the axiom of dependent choices, *J. Symb. Logic* **29**, 226.
[1965] Some applications of the notions of forcing and generic sets, *Fund. Math.* **56**, 325–345.

S. Feferman and A. Levy
[1963] Independence results in set theory by Cohen's method, *Notices Am. Math. Soc.* **10**, 593.

U. Felgner
[1969, 1970] Die Existenz wohlgeordneter, konfinaler Teilmengen in Ketten und das Auswahlaxiom, *Math. Z.* **111**, 221–232; correction: *ibid.* **115**, 392.
[1971a] *Models of* ZF *Set Theory*, Lecture Notes in Mathematics **223** (Springer, Berlin).
[1971b] Über das Ordnungstheorem, *Z. Math. Logik u. Grundl. Math.* **17**, 257–272.
[1971c] Comparison of the axioms of local and universal choice, *Fund. Math.* **71**, 43–62.

U. Felgner and T. Jech
[1973] Variants of the axiom of choice in set theory with atoms, *Fund. Math.*, to appear.

G. Fodor
[1966] On stationary sets and regressive functions, *Acta Sci. Math.* (*Szeged*) **27**, 105–110.

A. Fraenkel
[1922] Der Begriff "definit" und die Unabhängigkeit des Auswahlaxioms, *Sitzungsber. Preussischen Akad. Wiss., Math. Kl.*, 253–257.
[1925] Untersuchungen über die Grundlagen der Mengenlehre, *Math. Z.* **22**, 250–273.
[1928] Über die Ordnungsfähigkeit beliebiger Mengen, *Sitzungsber. Preussischen Akad. Wiss., Math. Kl.*, 90–91.
[1935] Sur l'axiome du choix, *L'Enseignement Math.* **34**, 32–51.
[1937] Über eine abgeschwächte Fassung des Auswahlaxioms, *J. Symb. Logic* **2**, 1–25.

R. Fraïssé
[1958] Un modèle définissant une théorie aberrante des ensembles où sont niés les axiomes du choix et d'extensionalité, *Publ. Sci. Univ. Alger, Sér.* A, **5**, 17–98.

H. Frampton
[1970] Zorn's Lemma, a motion picture (reviewed by H. Thompson in The New York Times, Sept. 14 (1970) 48 : 1).

R. J. Gauntt
[1967] Undefinability of cardinality, *Axiomatic Set Theory*, Proc. Symp. Pure Math., Univ. of California, Los Angeles, D. Scott, ed., **13** (2), to appear.
[1968] Some restricted versions of the axiom of choice, *Notices Am. Math. Soc.* **15**, 351
[1970] Axiom of choice for finite sets – A solution to a problem of Mostowski, *Notices Am. Math. Soc.* **17**, 454.

K. Gödel
[1938] The consistency of the axiom of choice and the generalized continuum hypothesis, *Proc. Natl. Acad. Sci. U.S.A.* **24**, 556–557.

[1939] Consistency-proof for the generalized continuum hypothesis, *Proc. Natl. Acad. Sci. U.S.A.* **25**, 220–224.

[1940] *The Consistency of the Axiom of Choice and of the Generalized Continuum Hypothesis*, Ann. Math. Studies **3** (Princeton Univ. Press, Princeton, N.J.).

P. HÁJEK

[1966] The consistency of the Church's alternatives, *Bull. Acad. Polon. Sci., Sér. Math.*, **14**, 423–430.

P. HÁJEK and P. VOPĚNKA

[1966] Some permutation submodels of the model ∇, *Bull. Acad. Polon. Sci., Sér. Math.*, **14**, 1–7.

J. D. HALPERN

[1962] Doctoral Dissertation, Univ. of California, Berkeley, Calif.

[1964] The independence of the axiom of choice from the Boolean prime ideal theorem, *Fund. Math.* **55**, 57–66.

[1972] On a question of Tarski and a maximal theorem of Kurepa, Pacific J. Math. **41**, 111–121.

J. D. HALPERN and P. E. HOWARD

[1970] Cardinals m such that $2m = m$, *Bull. Am. Math. Soc.* **76**, 487–490.

J. D. HALPERN and H. LÄUCHLI

[1966] A partition theorem, *Trans. Am. Math. Soc.* **124**, 360–367.

J. D. HALPERN and A. LEVY

[1964] The ordering theorem does not imply the axiom of choice, *Notices Am. Math. Soc.* **11**, 56.

[1967] The Boolean prime ideal theorem does not imply the axiom of choice, *Axiomatic Set Theory*, Proc. Symp. Pure Math., Univ. of California, Los Angeles, D. Scott, ed., **13** (1), 83–134.

F. HAUSDORFF

[1914] *Grundzüge der Mengenlehre* (Leipzig).

L. HENKIN

[1954] Metamathematical theorems equivalent to the prime ideal theorems for Boolean algebras, *Bull. Am. Math. Soc.* **60**, 388.

M. JAEGERMANN

[1965] The axiom of choice and two definitions of continuity, *Bull. Acad. Polon. Sci., Sér. Math.*, **13**, 699–704.

T. JECH

[1966a] On ordering of cardinalities, *Bull. Acad. Polon. Sci., Sér. Math.*, **14**, 293–296.

[1966b] On cardinals and their successors, *Bull. Acad. Polon. Sci., Sér. Math.*, **14**, 533–537.

[1966c] Interdependence of weakened forms of the axiom of choice, *Comment. Math. Univ. Carolinae* **7**, 359–371.

[1966d] Axiom výběru, Doctoral Dissertation, Charles University, Prague.

[1967] On models for set theory without AC, *Axiomatic Set Theory*, Proc. Symp. Pure Math., Univ. of California, Los Angeles, D. Scott, ed., **13** (1), 135–141.

[1968a] Eine Bemerkung zum Auswahlaxiom, *Časopis Pěst. Math.* **93**, 30–31.

[1968b] ω_1 can be measurable, *Israel J. Math.* **6**, 363–367.

[1971] *Lectures in Set Theory with Particular Emphasis on the Method of Forcing*, Lecture Notes in Math. **217** (Springer, Berlin).

T. JECH and A. SOCHOR

[1966a] On θ-model of the set theory, *Bull. Acad. Polon. Sci., Sér. Math.*, **14**, 297–303.

[1966b] Applications of the θ-model, *Bull. Acad. Polon. Sci., Sér. Math.*, **14**, 351–355.

R. B. JENSEN

[1967] Consistency results for ZF, *Notices Am. Math. Soc.* **14**, 137.

A. S. JESENIN-VOL'PIN

[1954] The unprovability of Suslin's hypothesis without the aid of the axiom of choice, *Dokl. Akad. Nauk. SSSR* **96**, 9–12 (in Russian).

B. JÓNSSON

[1960] Homogeneous universal relational systems, *Math. Scand.* **8**, 137–142.

J. L. KELLEY

[1950] The Tychonoff product theorem implies the axiom of choice, *Fund. Math.* **37**, 75–76.

W. KINNA and K. WAGNER

[1955] Über eine Abschwächung des Auswahlpostulates, *Fund. Math.* **42**, 75–82.

E. M. KLEINBERG

[1969] The independence of Ramsey's Theorem, *J. Symb. Logic* **34**, 205–206.

[1970] Strong partition properties for infinite cardinals, *J. Symb. Logic* **35**, 410–428.

C. KURATOWSKI

[1922] Une méthode d'élimination des nombres transfinis des raisonnements mathématiques, *Fund. Math.* **3**, 76–108.

D. KUREPA

[1953] Über das Auswahlaxiom, *Math. Ann.* **126**, 381–384.

H. LÄUCHLI

[1963] Auswahlaxiom in der Algebra, *Comment. Math. Helv.* **37**, 1–18.

[1964] The independence of the ordering principle from a restricted axiom of choice, *Fund. Math.* **54**, 31–43.

[1971] Coloring infinite graphs and the Boolean prime ideal theorem, *Israel J. Math.* **9**, 422–429.

A. LEVY

[1957] Indépendence conditionnelle de $V = L$, *C. R. Acad. Sci. Paris* **245**, 1582–1583.

[1958] The independence of various definitions of finiteness, *Fund. Math.* **46**, 1–13.

[1960] On models of set theory with urelements, *Bull. Acad. Polon. Sci., Sér. Math.* **8**, 463–465.

[1961] Comparing the axioms of local and universal choice, *Essays on the Foundations of Mathematics*, Y. Bar-Hillel, ed. (Jerusalem) 83–90.

[1962] Axioms of multiple choice, *Fund. Math.* **50**, 475–483.

[1963a] Remarks on a paper by J. Mycielski, *Acta Math. Acad. Sci. Hung.* **14**, 125–130.

[1963b] The Fraenkel-Mostowski method for independence proofs in set theory, *The Theory of Models*, J. Addison, L. Henkin and A. Tarski, eds. (North-Holland, Amsterdam, 1965) 221–228.

[1964] The interdependence of certain consequences of the axiom of choice, *Fund. Math.* **54**, 135–157.

[1967] On the logical complexity of several axioms of set theory, *Axiomatic Set Theory*, Proc. Symp. Pure Math., Univ. of California, Los Angeles, D. Scott, ed., **13** (1), 219–230.

[1969] The definability of cardinal numbers, *Foundations of Math.*, J. Bulloff, ed. (Springer, Berlin) 15–38.

A. LINDENBAUM and A. MOSTOWSKI

[1938] Über die Unabhängigkeit des Auswahlaxioms und einiger seiner Folgerungen, *C. R. Soc. Sci. Varsovie, Cl. III*, **31**, 27–32.

A. LINDENBAUM and A. TARSKI

[1926] Communications sur les recherches de la théorie des ensembles, *C.R. Soc. Sci. Varsovie, Cl.* III, **19**, 299–330.

J. ŁOŚ and C. RYLL-NARDZEWSKI

[1951] On the application of Tychonoff's theorem in mathematical proofs, *Fund. Math.* **38**, 233–237.

[1955] Effectiveness of the representation theory for Boolean algebras, *Fund. Math.* **41**, 49–56.

W. A. J. LUXEMBURG

[1962] Two applications of the method of construction by ultrapowers to analysis, *Bull. Am. Math. Soc.* **68**, 416–419.

[1969] Reduced powers of the real number system and equivalents of the Hahn–Banach extension theorem, *Intern. Symp. on the Applications of Model Theory*, (Holt, Rinehart and Winston, Toronto, Ont.) 123–137.

E. MARCZEWSKI

[1930] Sur l'extension de l'ordre partiel, *Fund. Math.* **16**, 386–389.

W. MAREK

[1966] A remark on independence proofs, *Bull. Acad. Polon. Sci., Sér. Math.*, **14**, 543–545.

W. MAREK and J. ONYSZKIEWICZ

[1966a] Representation of a partial ordering in cardinals of model of Zermelo–Fraenkel set theory, I, II, *Bull. Acad. Polon. Sci., Sér. Math.*, **14**, 357–358, 479–481.

[1966b] Generalized Dedekind numbers, *Bull. Acad. Polon. Sci., Sér. Math.*, **14**, 483–487.

D. A. MARTIN

[1968] The axiom of determinateness and reduction principles in the analytical hierarchy, *Bull. Am. Math. Soc.* **74**, 687–689.

A. R. D. MATHIAS

[1967] The order extension principle, to appear.

[1968] On an generalization of Ramsey's Theorem, *Notices Am. Math. Soc.* **15**, 931.

E. MENDELSON

[1948] The axiom of fundierung and the axiom of choice, *Arch. Math. Logik u. Grundl.forsch.* **4**, 65–70.

[1956] The independence of a weak axiom of choice, *J. Symb. Logic* **21**, 350–366.

G. P. MONRO

[1972] Models of ZF with the same set of sets of ordinals, *Notices Am. Math. Soc.*, **19**, 534.

R. MONTAGUE

[1961] Fraenkel's addition to the axioms of Zermelo, *Essays on the Foundations of Mathematics*, Y. Bar-Hillel, ed. (Jerusalem) 91–119.

D. B. MORRIS

[1969] Choice and cofinal well-ordered subsets, *Notices Am. Math. Soc.* **16**, 1088.

[1970] A model of ZF which cannot be extended to a model of ZFC without adding ordinals, *Notices Am. Math. Soc.* **17**, 577.

A. MOSTOWSKI

[1938] Über den Begriff einer endlichen Menge, *C.R. Soc. Sci. Varsovie, Cl.* III, **31**, 13–20.

[1939] Über die Unabhängigkeit des Wohlordnungssatzes vom Ordnungsprinzip, *Fund. Math.* **32**, 201–252.

[1945] Axiom of choice for finite sets, *Fund. Math.* **33**, 137–168.

[1948] On the principle of dependent choices, *Fund. Math.* **35**, 127–130.

[1949] An undecidable arithmetical statement, *Fund. Math.* **36**, 143–164.
[1958] On a problem of W. Kinna and K. Wagner, *Coll. Math.* **6**, 207–208.

S. Mrówka
[1955] On the ideals' extension theorem and its equivalence to the axiom of choice, *Fund. Math.* **43**, 46–49; **46**, 165–166.

J. Mycielski
[1961] Some remarks and problems on the colouring of infinite graphs and the theorem of Kuratowski, *Acta Math. Acad. Sci. Hung.* **12**, 125–129; **18**, 339–340.
[1964a] Two remarks on Tychonoff's product theorem, *Bull. Acad. Polon. Sci., Sér. Math.*, **8**, 439–441.
[1964b] On the axiom of determinateness, *Fund. Math.* **53**, 205–224.

J. Mycielski and S. Świerczkowski
[1964] On the Lebesgue measurability and the axiom of determinateness, *Fund. Math.* **54**, 67–71.

J. Myhill and D. Scott
[1967] Ordinal definability, *Axiomatic Set Theory*, Proc. Symp. Pure Math., Univ. of California, Los Angeles, D. Scott, ed., **13** (1), 271–278.

D. Pincus
[1968] Comparison of independence results in Mostowski's system (G) and ZF set theory, *Notices Am. Math. Soc.* **15**, 234.
[1969] Individuals in Zermelo–Fraenkel set theory, Doctoral Dissertation, Harvard Univ.
[1971] Support structures for the axiom of choice, *J. Symb. Logic* **36**, 28–38.
[1972a] Zermelo–Fraenkel consistencies by Fraenkel–Mostowski methods, *J. Symb. Logic* **37**, 721–743.
[1972b] Independence of the prime ideal theorem from the Hahn–Banach theorem, *Bull. Am. Math. Soc.* **78**, 766–770.
[1973a] On the independence of the Kinna–Wagner Principle, *Z. Math. Logik*, to appear.
[1973b] The strength of the Hahn–Banach Theorem, *Proc.* 1972 *Symp. on Nonstandard Analysis*, Lecture Notes in Math. (Springer, Berlin).

J. M. Plotkin
[1969] Generic embeddings, *J. Symb. Logic* **34**, 388–394.

R. Rado
[1949] Axiomatic treatment of rank in infinite sets. *Canad. J. Math.* **1**, 337–343.
[1954] Direct decomposition of partitions, *J. London Math. Soc.* **29**, 71–83.

F. P. Ramsey
[1929] On a problem of formal logic, *Proc. London Math. Soc., Ser.* 2, **30**, 264–286.

G. Rousseau
[1965] Note on a generalization of the Boolean ideal theorem equivalent to the axiom of choice, *Bull. Acad. Polon. Sci., Sér. Math.*, **13**, 521–522.

H. Rubin and J. Rubin
[1963] *Equivalents of the Axiom of Choice* (North-Holland, Amsterdam).

H. Rubin and D. Scott
[1954] Some topological theorems equivalent to the Boolean prime ideal theorem, *Bull. Am. Math. Soc.* **60**, 389.

B. Russel
[1906] On some difficulties in the theory of transfinite numbers and order types, *Proc. London Math. Soc., Ser.* 2, **4**, 29–53.

G. SAGEEV

[1973] An independence result concerning the axiom of choice, Preliminary report, *Notices Am. Math. Soc.* **20**, 22.

D. SCOTT

[1954a] The theorem on maximal ideals in lattices and the axiom of choice, *Bull. Am. Math. Soc.* **60**, 83.

[1954b] Prime ideal theorems for rings, lattices, and Boolean algebras, *Bull. Am. Math. Soc.* **60**, 390.

[1961] Measurable cardinals and constructible sets, *Bull. Acad. Polon. Sci., Sér. Math.*, **9**, 521–524.

D. SCOTT and R. M. SOLOVAY

[1967] Boolean valued models for set theory, to appear.

J. SHEPHERDSON

[1951] Inner models for set theory I, *J. Symb. Logic* **16**, 161–190.

J. R. SHOENFIELD

[1955] The independence of the axiom of choice, *J. Symb. Logic* **20**, 202.

W. SIERPIŃSKI

[1918] L'axiome de M. Zermelo et son rôle dans la théorie des ensembles et l'analyse *Bull. Acad. Sci. Cracovie, Cl. Sci. Math., Sér.* A, 97–152.

[1922] Sur l'égalité $2m = 2n$ pour les nombres cardinaux, *Fund. Math.* **3**, 1–6.

[1938] Fonctions additives non complètement additives et fonctions non mesurables, *Fund. Math.* **30**, 96–99.

[1947] L'hypothèse généralisée du continu et l'axiome du choix, *Fund. Math.* **34**, 1–5.

[1958] *Cardinal and Ordinal Numbers* (Warsaw).

R. M. SOLOVAY

[1965] The measure problem, *Notices Am. Math. Soc.* **12**, 217.

[1967a] Real-valued measurable cardinals, *Axiomatic Set Theory*, Proc. Symp. Pure Math., Univ. of California, Los Angeles, D. Scott, ed., **13** (1), 397–428.

[1967b] Measurable cardinals and the axiom of determinateness, mimeographed notes.

[1970] A model of set theory in which every set of reals is Lebesgue measurable, *Ann. Math.* **92**, 1–56.

E. SPECKER

[1954] Verallgemeinerte Kontinuumshypothese und Auswahlaxiom, *Archiv Math. (Basel)* **5**, 332–337.

[1957] Zur Axiomatik der Mengenlehre (Fundierungs- und Auswahlaxiom). *Z. Math. Logik u. Grundl. Math.* **3**, 173–210.

M. H. STONE

[1936] The theory of representations for Boolean algebras, *Trans. Am. Math. Soc.* **40**, 37–111.

W. SZMIELEW

[1947] On choices from finite sets, *Fund. Math.* **34**, 75–80.

M. TAKAHASHI

[1967] On incomparable cardinals, *Comment. Math. Univ. St. Pauli* **16**, 129–142.

A. TARSKI

[1924a] Sur quelques théorèmes qui équivalent à l'axiome du choix, *Fund. Math.* **5**, 147–154.

[1924b] Sur les ensembles finis, *Fund. Math.* **6**, 45–95.

[1948] Axiomatic and algebraic aspects of two theorems on sums of cardinals, *Fund. Math.* **35**, 79–104.

[1949] Cancellation laws in the arithmetic of cardinals, *Fund. Math.* **36,** 77–99.
[1954a] Prime ideal theorems . . ., *Bull. Amer. Math. Soc.* **60,** 390–391.
[1954b] Theorems on the existence of successors of cardinals, and the axiom of choice, *Indag. Math.* **16,** 26–32.
[1965] On the existence of large sets of Dedekind cardinals, *Notices Am. Math. Soc.* **12,** 719.

O. Teichmüller
[1939] Braucht der Algebraiker das Auswahlaxiom?, *Deutsche Math.* **4,** 567–577.

J. K. Truss
[1970] Finite versions of the axiom of choice, *Notices Am. Math. Soc.* **17,** 577, 694.
[1973] On successors in cardinal arithmetic, to appear.

J. W. Tukey
[1940] *Convergence and Uniformity in Topology,* Ann. Math. Studies **2** (Princeton Univ. Press, Princeton, N.J.).

S. Ulam
[1930] Zur Masstheorie in der allgemeinen Mengenlehre, Fund. Math. **16,** 140–150.

R. Vaught
[1952] On the equivalence of the axiom of choice and a maximal principle, *Bull. Am. Math. Soc.* **58,** 66.

G. Vitali
[1905] *Sul Problema della Mesura dei Gruppi di Punti di una Retta* (Bologna).

P. Vopěnka
[1967] General theory of ∇-models, *Comment. Math. Univ. Carolinae* **8,** 145–170.

P. Vopěnka and B. Balcar
[1967] On complete models of the set theory, *Bull. Acad. Polon. Sci., Sér. Math.,* **15,** 839–841.

P. Vopěnka and P. Hájek
[1965] Permutation submodels of the model ∇, *Bull. Acad. Polon. Sci., Sér. Math.,* **13,** 611–614.
[1972] *The Theory of Semisets* (North-Holland, Amsterdam).

L. E. Ward, Jr.
[1962] A weak Tychonoff theorem and the axiom of choice, *Proc. Am. Math. Soc.* **13,** 757–758.

B. Węglorz
[1969] A model of set theory *S* over a given Boolean algebra, *Bull. Acad. Polon. Sci., Sér. Math.,* **17,** 201–202.

K. Wiśniewski
[1967] Note on the axiom of choice for finite sets, *Bull. Acad. Polon. Sci., Sér. Math.,* **15,** 373–375 (in Russian).
[1968] Weakened forms of the axiom of choice for finite sets, *Bull Acad. Polon. Sci., Sér. Math.* **16,** 615–620.

E. Zermelo
[1904] Beweis dass jede Menge wohlgeordnet werden kann, *Math. Ann.* **59,** 514–516.

M. Zorn
[1935] A remark on method in transfinite algebra, *Bull. Am. Math. Soc.* **41,** 667–670.

M. M. Zuckerman

[1967] Multiple choice axioms, *Axiomatic Set Theory*, Proc. Symp. Pure Math., Univ. of California, Los Angeles, D. Scott, ed., **13** (1), 447–466.

[1969a] Unifying condition for implications among the axioms of choice for finite sets, *Pacific J. Math.* **28**, 233–242.

[1969b] Some theorems on the axioms of choice for finite sets, *Z. Math. Logik u. Grundl. Math.* **15**, 385–399.

[1969c] On choosing subsets of n-element sets, *Fund. Math.* **64**, 163–179.

[1969d] Number theoretic results related to the axioms of choice for finite sets, *Ill. J. Math.* **13**, 521–527.

AUTHOR INDEX

SUBJECT INDEX

LIST OF SYMBOLS

A	Antichain Principle
AC	Axiom of Choice
AC_κ	(p. 119)
ACF	AC for families of finite sets
ACW	AC for families of well orderable sets
C_n	AC for families of n-element sets
$cf(\kappa)$	cofinality of κ
$cl(X)$	closure under Gödel operations
(D)	(p. 116)
DC	Principle of Dependent Choices
DC_κ	(p. 119)
$dom(f)$	domain of f
$fix(x)$	$\{\pi: \pi y = y \text{ for all } y \in x\}$
glb	greatest lower bound
HOD	hereditarily ordinally definable
HS	hereditarily symmetric
L	constructible universe
lub	least upper bound
MC	Axiom of Multiple Choice
OD	ordinally definable
OEP	Order Extension Principle
On	class of all ordinals
P	(p. 133)
real	real number, set of natural numbers, function from ω to ω
$rng(f)$	range of f
$RO(P)$	regular open algebra
(S)	(pp. 111, 116)

sym(x)	$\{\pi: \pi x = x\}$
TC	transitive closure
$V = L$	Axiom of Constructibility
W_κ	(p. 119)
w.r.t.	with respect to
ZF	Zermelo–Fraenkel set theory
ZFA	ZF with atoms
$\aleph$	aleph
$\aleph(\mathfrak{p})$	Hartogs number
Δ, $\underline{\Delta}$	support relation
ΔC_α	diagonal intersection
$\kappa \to (\lambda)^2$	(p. 172)
$\kappa \to (\lambda)^\alpha$	(p. 174)
$\pi_2 \cdot \pi_1$	product of permutations: $\pi_2 \cdot \pi_1 x = \pi_2(\pi_1 x)$
$\prod$	(infinite) Boolean product
$\rho(x)$	rank in Boolean-valued models
$\sum$	(infinite) Boolean sum
ω	set of all natural numbers
$\wedge$	and
$\leftrightarrow$	if and only if
$\vec{x}$	$x_1, \ldots, x_n$
$\{u, v\}$	unordered pair
(u, v)	ordered pair
$\langle a_0, \ldots, a_k\rangle$	finite sequence
$\langle x_i : i \in I\rangle$	function on I
$\{x_n\}$	$\{x_n : n \in \omega\}$
$f : X \to Y$	function from X into Y
$f : x \mapsto y$	$f(x) = y$
$f''X$	$\{f(x) : x \in X\}$
$f \mid X$	restriction of f to X
${}^X Y$	set of all $f : X \to Y$
$\lvert X\rvert$	cardinality of X
$\mathscr{P}(X)$	set of all subsets of X
0	empty set, zero
$\mathscr{P}^\alpha(S)$	(p. 45)
V_α	sets of rank less than α

$\times\{S_i : i \in I\}$	Cartesian product
$[W]^r$	(p. 99)
$[A]^\alpha$	(pp. 173, 174)
$[H : K]$	index of subgroup
$[E]$	vector space generated by E
$F[x]$	ring of polynomials
R/I	quotient
$\mathscr{F}_1, \ldots, \mathscr{F}_8$	Gödel operations
$\mathscr{M} \models \phi$	ϕ holds in $\mathscr{M}$
$\mathscr{F}^{\mathscr{M}}(x)$	$\mathscr{F}$ performed inside $\mathscr{M}$
$\mathscr{M}^B$	Boolean valued model
$[\![\phi]\!]$	Boolean value
$\{x, y\}^B$, $(x, y)^B$	Boolean pairs
$\check{x}$	embedding of $\mathscr{M}$ into $\mathscr{M}^B$
$p \Vdash \phi$	p forces ϕ
$p{:}e$	(p. 72)
$s(p)$	(p. 73)
$\mathscr{M}[G]$	generic model
$i = i_G$	interpretation
$\underline{x}$	name for x

A CATALOG OF SELECTED

DOVER BOOKS

IN SCIENCE AND MATHEMATICS

Astronomy

BURNHAM'S CELESTIAL HANDBOOK, Robert Burnham, Jr. Thorough guide to the stars beyond our solar system. Exhaustive treatment. Alphabetical by constellation: Andromeda to Cetus in Vol. 1; Chamaeleon to Orion in Vol. 2; and Pavo to Vulpecula in Vol. 3. Hundreds of illustrations. Index in Vol. 3. 2,000pp. 6⅛ x 9¼.
Vol. I: 0-486-23567-X
Vol. II: 0-486-23568-8
Vol. III: 0-486-23673-0

EXPLORING THE MOON THROUGH BINOCULARS AND SMALL TELESCOPES, Ernest H. Cherrington, Jr. Informative, profusely illustrated guide to locating and identifying craters, rills, seas, mountains, other lunar features. Newly revised and updated with special section of new photos. Over 100 photos and diagrams. 240pp. 8¼ x 11. 0-486-24491-1

THE EXTRATERRESTRIAL LIFE DEBATE, 1750–1900, Michael J. Crowe. First detailed, scholarly study in English of the many ideas that developed from 1750 to 1900 regarding the existence of intelligent extraterrestrial life. Examines ideas of Kant, Herschel, Voltaire, Percival Lowell, many other scientists and thinkers. 16 illustrations. 704pp. 5⅜ x 8½. 0-486-40675-X

THEORIES OF THE WORLD FROM ANTIQUITY TO THE COPERNICAN REVOLUTION, Michael J. Crowe. Newly revised edition of an accessible, enlightening book recreates the change from an earth-centered to a sun-centered conception of the solar system. 242pp. 5⅜ x 8½. 0-486-41444-2

A HISTORY OF ASTRONOMY, A. Pannekoek. Well-balanced, carefully reasoned study covers such topics as Ptolemaic theory, work of Copernicus, Kepler, Newton, Eddington's work on stars, much more. Illustrated. References. 521pp. 5⅜ x 8½.
0-486-65994-1

A COMPLETE MANUAL OF AMATEUR ASTRONOMY: TOOLS AND TECHNIQUES FOR ASTRONOMICAL OBSERVATIONS, P. Clay Sherrod with Thomas L. Koed. Concise, highly readable book discusses: selecting, setting up and maintaining a telescope; amateur studies of the sun; lunar topography and occultations; observations of Mars, Jupiter, Saturn, the minor planets and the stars; an introduction to photoelectric photometry; more. 1981 ed. 124 figures. 25 halftones. 37 tables. 335pp. 6½ x 9¼. 0-486-40675-X

AMATEUR ASTRONOMER'S HANDBOOK, J. B. Sidgwick. Timeless, comprehensive coverage of telescopes, mirrors, lenses, mountings, telescope drives, micrometers, spectroscopes, more. 189 illustrations. 576pp. 5⅜ x 8¼. (Available in U.S. only.)
0-486-24034-7

STARS AND RELATIVITY, Ya. B. Zel'dovich and I. D. Novikov. Vol. 1 of *Relativistic Astrophysics* by famed Russian scientists. General relativity, properties of matter under astrophysical conditions, stars, and stellar systems. Deep physical insights, clear presentation. 1971 edition. References. 544pp. 5⅜ x 8¼. 0-486-69424-0

Chemistry

THE SCEPTICAL CHYMIST: THE CLASSIC 1661 TEXT, Robert Boyle. Boyle defines the term "element," asserting that all natural phenomena can be explained by the motion and organization of primary particles. 1911 ed. viii+232pp. 5⅜ x 8½. 0-486-42825-7

RADIOACTIVE SUBSTANCES, Marie Curie. Here is the celebrated scientist's doctoral thesis, the prelude to her receipt of the 1903 Nobel Prize. Curie discusses establishing atomic character of radioactivity found in compounds of uranium and thorium; extraction from pitchblende of polonium and radium; isolation of pure radium chloride; determination of atomic weight of radium; plus electric, photographic, luminous, heat, color effects of radioactivity. ii+94pp. 5⅜ x 8½. 0-486-42550-9

CHEMICAL MAGIC, Leonard A. Ford. Second Edition, Revised by E. Winston Grundmeier. Over 100 unusual stunts demonstrating cold fire, dust explosions, much more. Text explains scientific principles and stresses safety precautions. 128pp. 5⅜ x 8½. 0-486-67628-5

THE DEVELOPMENT OF MODERN CHEMISTRY, Aaron J. Ihde. Authoritative history of chemistry from ancient Greek theory to 20th-century innovation. Covers major chemists and their discoveries. 209 illustrations. 14 tables. Bibliographies. Indices. Appendices. 851pp. 5⅜ x 8½. 0-486-64235-6

CATALYSIS IN CHEMISTRY AND ENZYMOLOGY, William P. Jencks. Exceptionally clear coverage of mechanisms for catalysis, forces in aqueous solution, carbonyl- and acyl-group reactions, practical kinetics, more. 864pp. 5⅜ x 8½. 0-486-65460-5

ELEMENTS OF CHEMISTRY, Antoine Lavoisier. Monumental classic by founder of modern chemistry in remarkable reprint of rare 1790 Kerr translation. A must for every student of chemistry or the history of science. 539pp. 5⅜ x 8½. 0-486-64624-6

THE HISTORICAL BACKGROUND OF CHEMISTRY, Henry M. Leicester. Evolution of ideas, not individual biography. Concentrates on formulation of a coherent set of chemical laws. 260pp. 5⅜ x 8½. 0-486-61053-5

A SHORT HISTORY OF CHEMISTRY, J. R. Partington. Classic exposition explores origins of chemistry, alchemy, early medical chemistry, nature of atmosphere, theory of valency, laws and structure of atomic theory, much more. 428pp. 5⅜ x 8½. (Available in U.S. only.) 0-486-65977-1

GENERAL CHEMISTRY, Linus Pauling. Revised 3rd edition of classic first-year text by Nobel laureate. Atomic and molecular structure, quantum mechanics, statistical mechanics, thermodynamics correlated with descriptive chemistry. Problems. 992pp. 5⅜ x 8½. 0-486-65622-5

FROM ALCHEMY TO CHEMISTRY, John Read. Broad, humanistic treatment focuses on great figures of chemistry and ideas that revolutionized the science. 50 illustrations. 240pp. 5⅜ x 8½. 0-486-28690-8

Engineering

DE RE METALLICA, Georgius Agricola. The famous Hoover translation of greatest treatise on technological chemistry, engineering, geology, mining of early modern times (1556). All 289 original woodcuts. 638pp. 6¾ x 11. 0-486-60006-8

FUNDAMENTALS OF ASTRODYNAMICS, Roger Bate et al. Modern approach developed by U.S. Air Force Academy. Designed as a first course. Problems, exercises. Numerous illustrations. 455pp. 5⅜ x 8½. 0-486-60061-0

DYNAMICS OF FLUIDS IN POROUS MEDIA, Jacob Bear. For advanced students of ground water hydrology, soil mechanics and physics, drainage and irrigation engineering and more. 335 illustrations. Exercises, with answers. 784pp. 6⅛ x 9¼. 0-486-65675-6

THEORY OF VISCOELASTICITY (Second Edition), Richard M. Christensen. Complete consistent description of the linear theory of the viscoelastic behavior of materials. Problem-solving techniques discussed. 1982 edition. 29 figures. xiv+364pp. 6⅛ x 9¼. 0-486-42880-X

MECHANICS, J. P. Den Hartog. A classic introductory text or refresher. Hundreds of applications and design problems illuminate fundamentals of trusses, loaded beams and cables, etc. 334 answered problems. 462pp. 5⅜ x 8½. 0-486-60754-2

MECHANICAL VIBRATIONS, J. P. Den Hartog. Classic textbook offers lucid explanations and illustrative models, applying theories of vibrations to a variety of practical industrial engineering problems. Numerous figures. 233 problems, solutions. Appendix. Index. Preface. 436pp. 5⅜ x 8½. 0-486-64785-4

STRENGTH OF MATERIALS, J. P. Den Hartog. Full, clear treatment of basic material (tension, torsion, bending, etc.) plus advanced material on engineering methods, applications. 350 answered problems. 323pp. 5⅜ x 8½. 0-486-60755-0

A HISTORY OF MECHANICS, René Dugas. Monumental study of mechanical principles from antiquity to quantum mechanics. Contributions of ancient Greeks, Galileo, Leonardo, Kepler, Lagrange, many others. 671pp. 5⅜ x 8½. 0-486-65632-2

STABILITY THEORY AND ITS APPLICATIONS TO STRUCTURAL MECHANICS, Clive L. Dym. Self-contained text focuses on Koiter postbuckling analyses, with mathematical notions of stability of motion. Basing minimum energy principles for static stability upon dynamic concepts of stability of motion, it develops asymptotic buckling and postbuckling analyses from potential energy considerations, with applications to columns, plates, and arches. 1974 ed. 208pp. 5⅜ x 8½. 0-486-42541-X

METAL FATIGUE, N. E. Frost, K. J. Marsh, and L. P. Pook. Definitive, clearly written, and well-illustrated volume addresses all aspects of the subject, from the historical development of understanding metal fatigue to vital concepts of the cyclic stress that causes a crack to grow. Includes 7 appendixes. 544pp. 5⅜ x 8½. 0-486-40927-9

Mathematics

FUNCTIONAL ANALYSIS (Second Corrected Edition), George Bachman and Lawrence Narici. Excellent treatment of subject geared toward students with background in linear algebra, advanced calculus, physics and engineering. Text covers introduction to inner-product spaces, normed, metric spaces, and topological spaces; complete orthonormal sets, the Hahn-Banach Theorem and its consequences, and many other related subjects. 1966 ed. 544pp. 6⅛ x 9¼. 0-486-40251-7

ASYMPTOTIC EXPANSIONS OF INTEGRALS, Norman Bleistein & Richard A. Handelsman. Best introduction to important field with applications in a variety of scientific disciplines. New preface. Problems. Diagrams. Tables. Bibliography. Index. 448pp. 5⅜ x 8½. 0-486-65082-0

VECTOR AND TENSOR ANALYSIS WITH APPLICATIONS, A. I. Borisenko and I. E. Tarapov. Concise introduction. Worked-out problems, solutions, exercises. 257pp. 5⅝ x 8¼. 0-486-63833-2

AN INTRODUCTION TO ORDINARY DIFFERENTIAL EQUATIONS, Earl A. Coddington. A thorough and systematic first course in elementary differential equations for undergraduates in mathematics and science, with many exercises and problems (with answers). Index. 304pp. 5⅜ x 8½. 0-486-65942-9

FOURIER SERIES AND ORTHOGONAL FUNCTIONS, Harry F. Davis. An incisive text combining theory and practical example to introduce Fourier series, orthogonal functions and applications of the Fourier method to boundary-value problems. 570 exercises. Answers and notes. 416pp. 5⅜ x 8½. 0-486-65973-9

COMPUTABILITY AND UNSOLVABILITY, Martin Davis. Classic graduate-level introduction to theory of computability, usually referred to as theory of recurrent functions. New preface and appendix. 288pp. 5⅜ x 8½. 0-486-61471-9

ASYMPTOTIC METHODS IN ANALYSIS, N. G. de Bruijn. An inexpensive, comprehensive guide to asymptotic methods–the pioneering work that teaches by explaining worked examples in detail. Index. 224pp. 5⅜ x 8½ 0-486-64221-6

APPLIED COMPLEX VARIABLES, John W. Dettman. Step-by-step coverage of fundamentals of analytic function theory–plus lucid exposition of five important applications: Potential Theory; Ordinary Differential Equations; Fourier Transforms; Laplace Transforms; Asymptotic Expansions. 66 figures. Exercises at chapter ends. 512pp. 5⅜ x 8½. 0-486-64670-X

INTRODUCTION TO LINEAR ALGEBRA AND DIFFERENTIAL EQUATIONS, John W. Dettman. Excellent text covers complex numbers, determinants, orthonormal bases, Laplace transforms, much more. Exercises with solutions. Undergraduate level. 416pp. 5⅜ x 8½. 0-486-65191-6

RIEMANN'S ZETA FUNCTION, H. M. Edwards. Superb, high-level study of landmark 1859 publication entitled "On the Number of Primes Less Than a Given Magnitude" traces developments in mathematical theory that it inspired. xiv+315pp. 5⅜ x 8½. 0-486-41740-9

INTRODUCTORY REAL ANALYSIS, A.N. Kolmogorov, S. V. Fomin. Translated by Richard A. Silverman. Self-contained, evenly paced introduction to real and functional analysis. Some 350 problems. 403pp. 5⅜ x 8½. 0-486-61226-0

APPLIED ANALYSIS, Cornelius Lanczos. Classic work on analysis and design of finite processes for approximating solution of analytical problems. Algebraic equations, matrices, harmonic analysis, quadrature methods, much more. 559pp. 5⅜ x 8½. 0-486-65656-X

AN INTRODUCTION TO ALGEBRAIC STRUCTURES, Joseph Landin. Superb self-contained text covers "abstract algebra": sets and numbers, theory of groups, theory of rings, much more. Numerous well-chosen examples, exercises. 247pp. 5⅜ x 8½. 0-486-65940-2

QUALITATIVE THEORY OF DIFFERENTIAL EQUATIONS, V. V. Nemytskii and V.V. Stepanov. Classic graduate-level text by two prominent Soviet mathematicians covers classical differential equations as well as topological dynamics and ergodic theory. Bibliographies. 523pp. 5⅜ x 8½. 0-486-65954-2

THEORY OF MATRICES, Sam Perlis. Outstanding text covering rank, nonsingularity and inverses in connection with the development of canonical matrices under the relation of equivalence, and without the intervention of determinants. Includes exercises. 237pp. 5⅜ x 8½. 0-486-66810-X

INTRODUCTION TO ANALYSIS, Maxwell Rosenlicht. Unusually clear, accessible coverage of set theory, real number system, metric spaces, continuous functions, Riemann integration, multiple integrals, more. Wide range of problems. Undergraduate level. Bibliography. 254pp. 5⅜ x 8½. 0-486-65038-3

MODERN NONLINEAR EQUATIONS, Thomas L. Saaty. Emphasizes practical solution of problems; covers seven types of equations. ". . . a welcome contribution to the existing literature...."–*Math Reviews*. 490pp. 5⅜ x 8½. 0-486-64232-1

MATRICES AND LINEAR ALGEBRA, Hans Schneider and George Phillip Barker. Basic textbook covers theory of matrices and its applications to systems of linear equations and related topics such as determinants, eigenvalues and differential equations. Numerous exercises. 432pp. 5⅜ x 8½. 0-486-66014-1

LINEAR ALGEBRA, Georgi E. Shilov. Determinants, linear spaces, matrix algebras, similar topics. For advanced undergraduates, graduates. Silverman translation. 387pp. 5⅜ x 8½. 0-486-63518-X

ELEMENTS OF REAL ANALYSIS, David A. Sprecher. Classic text covers fundamental concepts, real number system, point sets, functions of a real variable, Fourier series, much more. Over 500 exercises. 352pp. 5⅜ x 8½. 0-486-65385-4

SET THEORY AND LOGIC, Robert R. Stoll. Lucid introduction to unified theory of mathematical concepts. Set theory and logic seen as tools for conceptual understanding of real number system. 496pp. 5⅜ x 8¼. 0-486-63829-4

Physics

OPTICAL RESONANCE AND TWO-LEVEL ATOMS, L. Allen and J. H. Eberly. Clear, comprehensive introduction to basic principles behind all quantum optical resonance phenomena. 53 illustrations. Preface. Index. 256pp. 5⅜ x 8½. 0-486-65533-4

QUANTUM THEORY, David Bohm. This advanced undergraduate-level text presents the quantum theory in terms of qualitative and imaginative concepts, followed by specific applications worked out in mathematical detail. Preface. Index. 655pp. 5⅜ x 8½. 0-486-65969-0

ATOMIC PHYSICS (8th EDITION), Max Born. Nobel laureate's lucid treatment of kinetic theory of gases, elementary particles, nuclear atom, wave-corpuscles, atomic structure and spectral lines, much more. Over 40 appendices, bibliography. 495pp. 5⅜ x 8½. 0-486-65984-4

A SOPHISTICATE'S PRIMER OF RELATIVITY, P. W. Bridgman. Geared toward readers already acquainted with special relativity, this book transcends the view of theory as a working tool to answer natural questions: What is a frame of reference? What is a "law of nature"? What is the role of the "observer"? Extensive treatment, written in terms accessible to those without a scientific background. 1983 ed. xlviii+172pp. 5⅜ x 8½. 0-486-42549-5

AN INTRODUCTION TO HAMILTONIAN OPTICS, H. A. Buchdahl. Detailed account of the Hamiltonian treatment of aberration theory in geometrical optics. Many classes of optical systems defined in terms of the symmetries they possess. Problems with detailed solutions. 1970 edition. xv + 360pp. 5⅜ x 8½. 0-486-67597-1

PRIMER OF QUANTUM MECHANICS, Marvin Chester. Introductory text examines the classical quantum bead on a track: its state and representations; operator eigenvalues; harmonic oscillator and bound bead in a symmetric force field; and bead in a spherical shell. Other topics include spin, matrices, and the structure of quantum mechanics; the simplest atom; indistinguishable particles; and stationary-state perturbation theory. 1992 ed. xiv+314pp. 6⅛ x 9¼. 0-486-42878-8

LECTURES ON QUANTUM MECHANICS, Paul A. M. Dirac. Four concise, brilliant lectures on mathematical methods in quantum mechanics from Nobel Prize-winning quantum pioneer build on idea of visualizing quantum theory through the use of classical mechanics. 96pp. 5⅜ x 8½. 0-486-41713-1

THIRTY YEARS THAT SHOOK PHYSICS: THE STORY OF QUANTUM THEORY, George Gamow. Lucid, accessible introduction to influential theory of energy and matter. Careful explanations of Dirac's anti-particles, Bohr's model of the atom, much more. 12 plates. Numerous drawings. 240pp. 5⅜ x 8½. 0-486-24895-X

ELECTRONIC STRUCTURE AND THE PROPERTIES OF SOLIDS: THE PHYSICS OF THE CHEMICAL BOND, Walter A. Harrison. Innovative text offers basic understanding of the electronic structure of covalent and ionic solids, simple metals, transition metals and their compounds. Problems. 1980 edition. 582pp. 6⅛ x 9¼. 0-486-66021-4